mechanical engineering

THE SOURCES OF INFORMATION

mechanical engineering

THE SOURCES OF INFORMATION

BERNARD HOUGHTON
FLA

SENIOR LECTURER
DEPARTMENT OF LIBRARY AND INFORMATION STUDIES
LIVERPOOL POLYTECHNIC

ARCHON BOOKS & CLIVE BINGLEY

FIRST PUBLISHED 1970 BY CLIVE BINGLEY LTD
THIS EDITION SIMULTANEOUSLY PUBLISHED IN THE USA
BY ARCHON BOOKS, THE SHOE STRING PRESS INC,
995 SHERMAN AVENUE, HAMDEN, CONNECTICUT 06514
PRINTED IN GREAT BRITAIN
208 01062 9

CONTENTS

INTRODUCTION

This work is offered as a practical guide to sources of information on mechanical engineering and related technologies, for use by practical engineers, engineering research workers and librarians and information officers working in libraries and information services serving mechanical engineers.

It is not intended as a bibliography of mechanical engineering literature or as a listing of titles, but as a map to help the engineer to find his way through the varying forms of literature. The approach has been to describe briefly the situations in which research is conducted in the United Kingdom and the United States and then to describe the various categories of publication which are published to document this research, indicating for each form its role and value to the engineer.

Periodicals and reference tools have been treated exhaustively and it is hoped that the chapters covering these forms will be of particular value both to engineers who wish to ascertain which are the chief publications covering their interests and also to librarians who wish to select material for new or developing collections.

No attempt has been made to cover individual textbooks or standard works. The rate of technical publication is such that any treatment of this material would have been rendered out-of-date on publication. The approach here has been to list the essential book reviewing and selection services, and thus guide the engineer who needs to choose a particular type of text to the most relevant current aid to selection.

My thanks are due to Mr T M Chesworth, Mr H L Houghton and Mr E Walshaw of the Department of Mechanical Marine and Production Engineering of the Liverpool Polytechnic and also to Mr B Yates, for their advice in the drawing up of the subject scope of this book.

Thanks are also due to the library and information staffs of the Production Engineering Research Association and the Machine Tool Industry Research Association, and to the staffs of the National Reference Library for Science and Invention, the National Lending Library for Science and Technology, the Library of the Institution of Mechanical Engineers and especially to the staffs of the Liverpool Technical Library, LADSIRLAC, and the Picton Reference Library for their assistance in giving me access to the materials covered in this volume.

BERNARD HOUGHTON

Liverpool
December 1969

CHAPTER 1

INTRODUCTION

The function of the engineer is the solution of problems, decisions must be made in the solution of these problems and to reach them the engineer has need of information. D G Ainley, of the National Gas Turbine Establishment, has pointed out that the information needs of the mechanical engineer can be grouped under four broad headings: [1]

i) Information on competitors' activities, including news of their current research projects, the quality of their products and their planned developments.

ii) Information on market trends.

iii) News of the location, objectives and results of national and international research.

iv) General engineering knowledge in the form of ' spin-off ' from developments in fields of engineering other than his own.

The self-confessed need of the mechanical engineer for information at different levels has been underlined in a recent report by Wood and Hamilton of the National Lending Library for Science and Technology, whose survey of the requirements of 2,500 members of the Institution of Mechanical Engineers indicates that within the week before completing a questionnaire for the survey, 18·6 percent of the sample had required exhaustive information on a single topic within their field, 68·8 percent had required everyday information and 27·5 percent had required information from outside the field of mechanical engineering. [2] Much of this information can be obtained from the literature of science and technology, which is growing at an ever increasing rate, but at present few engineers are equipped to make the best use of the literature. Surveys which have been made of the information gathering habits of engineers and scientists have in general indicated that workers in the more exact sciences tend to depend more heavily on the literature covering their subjects than do applied scientists and engineers. The most obvious explanation of this

9

situation is that the professional training of the engineer is not literature oriented.

The Wood & Hamilton report demonstrated that only 11 percent of mechanical engineers had received any training in the use of literature, but of this figure 91 percent had found the training valuable. On the other hand, 75 percent of the engineers who had received no training indicated that they would have welcomed it. An earlier survey undertaken by the American Institute of Chemical Engineers pointed out that engineers are often unaware of sources of information on their subjects, and that there was a pressing need to introduce courses of instruction in the use of the literature into the professional training curricula of engineers.[3] Mr Ainley, speaking at a conference on the engineering information problem organised by the Institution of Mechanical Engineers, likened groups of engineers competing with other groups to ' teams of athletes competing in a nightmare Olympic games run in a dense fog. Competitors were never sure of their position.'

All the major surveys of the information gathering habits of engineers and scientists have indicated that verbal communication with a colleague was the most preferred method of gaining information. This method is convenient and has the advantage of providing opportunity for discussion, but the information thus available to the engineer is circumscribed by the experience of his immediate colleagues. The information available in the literature represents the theories, ideas, opinions of thousands of engineers, and the solutions to the problems by which they have been confronted. Requests for exhaustive information on a particular topic can only be answered satisfactorily by referring to the literature. It is not claimed that mechanical engineers *never* refer to the literature, but it has been demonstrated that their reading habits are often extremely parochial. Wood and Hamilton discovered that although the average engineer consults between five and ten journals regularly, he rarely consults a journal published outside the UK, and hardly ever refers to the abstracting and indexing services which are the key to the world's literature. The fact that the literature is not being used to any great extent by engineers means that extensive duplication of effort is occurring. Mr Martyn, of Aslib, estimated that in 1962 £6 million of the estimated £640 million spent in the UK on research and development had been expended unnecessarily because of needless duplication of work through published information not being discovered earlier. This is tantamount to paying 750 scientists each year to do nothing.[4]

Although no guide specifically to sources of information in mechanical engineering has hitherto been published, there are two useful general guides to sources of information in science and engineering, and also several more detailed guides to particular disciplines related to mechanical engineering, to which the mechanical engineer can refer when seeking information outside his immediate subject field. The two general guides are:

Jenkins, F B: *Science reference sources.* Champaign, Ill, Illini Union Bookstore, fourth edition 1965; a simple listing of basic reference sources without description or annotation. Categories included in the eleven subject groups, one of which is engineering, include guides to the literature, indexes, abstracts, reviews, dictionaries, encyclopedias, handbooks, tables, book selection aids, etc.

Malinowsky, Harold Robert: *Science and engineering reference sources: a guide for students and librarians.* Rochester, NJ, Libraries Unlimited, 1967; a more ambitious guide, with preliminary sections on literature searching techniques and the various forms of scientific and technical documents. The subject sections, covering mathematics, physics, engineering etc, are subdivided by such headings as indexes, abstracts, handbooks, etc.

The only specific guide to engineering literature is now very much out of date and consequently is of little value: Dalton, B H: *Sources of engineering information.* Berkeley, University of California Press, 1948.

Recent years have seen the publication of a number of comprehensive guides to particular subject fields, including:

AEROSPACE TECHNOLOGY

Fry, Bernard Mitchell & Mohrhardt, Foster E (eds): *A guide to sources in space science and technology.* New York, Interscience, 1963.

ATOMIC ENERGY

Anthony, L J: *Sources of information on atomic energy.* Oxford, Pergamon Press, 1966.

CHEMISTRY

Burman, C R: *How to find out in chemistry.* Oxford, Pergamon Press, second edition, 1966.

ELECTRICAL ENGINEERING

Burkett, Jack & Plumb, Philip: *How to find out in electrical engineering: a guide to sources of information.* Oxford, Pergamon Press, 1967.

MATHEMATICS

Parke, N G: *Guide to the literature of mathematics and physics, including related works on engineering science.* New York, Dover Publications, second edition 1958.

METALLURGY

Gibson, Eleanor B & Tapia, Elizabeth W (eds): *Guide to metallurgical information.* New York, Special Libraries Association, second edition 1965.

PHYSICS

Yates, Bryan: *How to find out about physics: a guide to sources of information.* Oxford, Pergamon Press, 1965.

REFRIGERATION ENGINEERING

Codlin, E M: *Cryogenics and refrigeration: a bibliographical guide.* London, Macdonald and Co, 1968.

Before considering the various forms of primary publication which document advances in engineering, and the secondary sources such as abstracting and indexing services, which exercise control over the primary sources and enable scientists and engineers to ascertain what work has already been carried out in a particular field, it is pertinent to consider the development of scientific research and its present organisation. The organisations pursuing research are both the main producers of the primary publications and the chief sponsors of abstracting and indexing services. The so-called ' information explosion ', with its resulting mass of publication, is a direct consequence of the massive increase in industrial research effort.

Research is trained observation and enquiry directed towards any department of knowledge with a view to the discovery of new information. It has been classified by the 1947 US President's Research Board into basic and applied research.[5] The board further subdivides basic research into fundamental research and background research. Fundamental research was defined as the using of ' theoretical analysis, exploration or experimentation directed to the extension of knowledge of general principles governing natural and social phenomena '. Background research is the ' systematic observation, collection, organisation and presentation of facts, using known principles, to reach objectives that are clearly defined before research is undertaken, to promote a foundation for subsequent research, or to provide standard reference data '. Applied research is ' the extension of basic research to the determination of generally accepted principles with a view to scientific application, generally involving the discovery of a specified novel product, process, technique or device '.

Between research and production comes the development stage, which is the final phase in the effort to produce social and economic benefits from research. Development is concerned with the exploitation of rather than the addition to knowledge, and involves the solution of production problems and the assessment of commercial potential. Development encompasses the adaptation of research findings to experimental demonstration and the experimental production and testing of models, equipment, materials, processes or procedures. The term ' industrial research ' is commonly used to cover both applied research and development.

In order to view the present state of organised scientific and technological research in the correct perspective it is necessary to examine both the working conditions which have been conducive to scientific innovation and the past development of research. Four divergent streams of research can be identified:

i) Discovery related to the exercise of a craft where experience in an art or chance observation during a traditional process may have suggested improvement in procedure or design.

ii) Experiments conducted by wealthy amateurs for diversion which have yielded unforeseen but important results.

iii) Discovery related to teaching as evidenced in academic research where discoveries are made or theories developed by academics through reflection on the principles which they teach.

iv) Discovery through professional research.

Up to the middle of the nineteenth century the main advances in both science and technology were effected chiefly by the craftsman or the wealthy amateur. The significance of academic research was not appreciated, and the profession of research scientist was not established in this country until the second decade of the present century. This was the age of the practical man; the links between science and technology were extremely tenuous. In the early years of the last century lack of academic training was in some measure compensated for by the simple nature of applicable science. It was possible for the amateur to be accepted into the higher ranks of both science and industry. Even towards the end of the century major inventions were still coming from a genre of inventors who, lacking an academic background, gleaned what science they needed from the few available textbooks and their own hard practical experience in workshops. There existed what would be regarded today as an alarming delay between a scientific discovery and its industrial application. It is indeed remarkable that the development of the steam engine proceeded so far without any clear understanding of the true nature of heat. The first and second laws of thermodynamics were enunciated by Joule in 1847, and Clausius and Lord Kelvin in 1851, but their significance in the design of the steam engine was not fully appreciated until decades later. Inventions were often based on ' serendipity ', or on ingenious intuition rather than on a sound understanding of underlying scientific principles. Two examples from the steel industry will serve to illustrate this. Joseph Hall (1789-1862), who was responsible for the development of the wet puddling process, was originally employed in a ' Black Country ' ironworks. In 1816, noting the widespread wastage necessitated in contemporary puddling processes, he decided to see what would happen when he attempted to work scraps of iron slag which had collected in the water tanks used for cooling the puddlers' tools. He had little conception of the principles of the ensuing reaction, which produced the best wrought iron he had ever seen. His process was soon to supercede all other processes for the production of wrought iron and was to make him a large fortune.

The Bessemer process of steel production was developed by Henry Bessemer (1813-1898), a man with little knowledge of iron metallurgy but possessed of immense natural inquisitiveness and acute powers of observation. While experimenting on ironmaking for gun barrels he observed that some pieces of pig-iron which remained unmelted at the side of his furnace had become decarburized, presumably by exposure to air passing through the furnace. This observation led him to try passing air through molten iron, resulting in the development of his historic Bessemer converter.

The gradual progression from craftsman or amateur working in isolation towards cooperative research effort was initially almost imperceptible and has only proceeded with any great momentum since the first world war. The nineteenth century has been called by Derek de Solla Price the age of the 'little science', exemplified by the individual scientist working in isolation, whereas the twentieth century is the age of the 'big science', characterised by the cooperative research team supported by large research funds subscribed by both government and industry.[6] With the development of cooperative research the delay between the discovery of a principle and its industrial application is rapidly decreasing. The engineer is today responsible for translating the results of scientific research into industrial hardware.

REFERENCES

1 Ainley, D G: 'Problems of information retrieval' *Chartered mechanical engineer*, *12*(4), April 1965, p. 211.

2 Wood, D N & Hamilton, D R L: *The information requirements of mechanical engineers*. London, Library Association, 1967.

3 Davis, R A: 'How engineers use literature' *Chemical engineering progress*, *61*(3), March 1965.

4 Cook, Sir James: 'The science information problem' *Advancement of science*, *23*(112) October 1966, p 305.

5 United States President's Scientific Research Board: *Science and public policy*. Washington DC, 1947.

6 De Solla Price, Derek J: *Little science, big science*. New York, Columbia University Press, 1963.

CHAPTER 2

RESEARCH IN THE UNITED KINGDOM
PROFESSIONAL SOCIETIES AND INSTITUTIONS, UNIVERSITIES

SOCIETIES AND INSTITUTIONS: The beginnings of scientific research in the United Kingdom can be traced to the foundation of the Royal Society of London in 1660. The main objects of the society were to enlarge knowledge by experiment and observation. The society was independent of the state and was financed through the subscriptions of its Fellows, although it did receive from time to time small amounts of money from the Treasury to assist in the financing of scientific research projects. In 1664 eight committees were established to examine specific areas of science with a view to research, one of the most popular of these being the committee for the study of mechanical questions, which attracted sixty nine members. An early attempt was made then to apply the discoveries of science to industrial use. Writing in 1664 Robert Hooke observed ' many of their members are men of commerce and traffick, which is a good omen that their attempts will bring philosophie from words to action '.[1] In his history of the Society published in 1667 Thomas Sprat notes ' the genious of experimenting is so much dispersed . . . that all places and corners are busy and warm about the work '.[2]

The Royal Society gave actual financial support for research by its Fellows, examples of its patronage being the help given to John Smeaton in 1759 to enable him to visit Flanders and Holland to pursue his study of windmills, and the financial aid given to Joule a century later to assist his research work on low-temperature refrigeration. The *Philosophical transactions of the Royal Society*, first published in 1665, constitute a record of the experimental work carried out by the Fellows, and are the first example of a scientific periodical to be published in the English language.

The society no longer undertakes experimental methods, but rather tenders advice to the government on scientific questions of all kinds, its public responsibilities being:

    i) Parliamentary grants in aid of scientific investigations—the society was first charged with the responsibility of administering government grants for research in 1849 when the sum of £1,000 per annum was made available. In 1967 £144,000 was distributed by the society, the grants being apportioned by a series of sectional committees, each dealing with a particular area of pure science or technology.

ii) Parliamentary grants in aid of international research associations and scientific congresses. Public funds are administered by the Royal Society to cover the expenses of UK delegates to international scientific meetings, and also to enable British scientists to gain experience in new techniques at foreign laboratories.

iii) Grants in support of research professorships at the universities.

iv) Grants in aid of scientific publication.

v) Joint responsibility with the Ministry of Technology for the management of the scientific policy of the National Physical Laboratory.

The Royal Society did not, however, maintain its own research laboratories; indeed, what was virtually the only research laboratory in the United Kingdom in the first half of the nineteenth century was maintained at the Royal Institution, established in 1799 through the foresight of Count Rumford as ' a public institution for the diffusing of knowledge and facilitating the general introduction of useful mechanical inventions and improvements and for teaching by courses of philosophical lectures the applications of science to the common purposes of life '. The Royal Institution was supported by the bequests of Rumford, the subscriptions of its members, and public donation, and from 1799 until the middle of the following century an unbroken sequence of important scientific discoveries emanated from its laboratory, many of these being the fundamental ideas from which modern scientific and technological progress has stemmed. In addition to the pursuit of research, the Royal Institution promoted the awareness of scientific principles by organising courses of public lectures which in the nineteenth century attracted large and enthusiastic audiences. The institution's current activities are still mainly concerned with research and lectures. Research is now carried out by a team of twenty five scientists under the supervision of a scientific director. One of the main subject areas of the work is deformation in metals.

By the late eighteenth century the prestige of the Royal Society had waned and the society had degenerated into a London club with half its Fellows being non-scientists. A contemporary critic of the society maintained that ' as a body it scarcely labours itself '. Charles Babbage criticised the society for failing to live up to its seventeenth century aims ' to improve the knowledge of natural things, and all useful arts, manufactures and mechanick practices, engynes and inventions and experiments '. Babbage insisted that if progress was to be made in science and technology the old interested amateur tradition in scientific research was no longer adequate and that a new association was needed to promote science. In 1831, as a result of his efforts, the British Association for the Advancement of Science was formed ' to give stronger impulse to and more systematic direction to scientific enquiry, to obtain a greater degree of national attention to the objects of science and a removal of the disadvantages which impede its progress '. For the remainder of that century the British Association was to be the spearhead of scientific advance in the United Kingdom. The ideas of both the Royal Society and the British Association had been anticipated by Francis Bacon, who had envisaged in his ' New Atlantis ' first published in 1626, a national academy for the advancement of science which would ' circulate the diverse principal cities ', as the British Association was to do by holding its annual meetings in different locations throughout the country some three centuries later.

Babbage did not succeed in eliciting government financial support for scientific research, but the British Association was, from its own funds, able to make small sums of money available for specific projects. In the first hundred years of its existence £92,000 was spent in assistance to research, the physical sciences—from which sprang the important engineering applications of the century—receiving a third of this sum.

Today the British Association is no longer concerned with the financing of research; its present efforts are mainly directed towards promoting a public relations service for the wider understanding of science and the cross fertilization of ideas between scientists and engineers working in different disciplines.

The progression from ' umbrella ' societies covering all disciplines to specialised scientific societies was symptomatic of the transition from interested amateur to specialist. As early as 1767 Priestley had declared ' at present there are, in different countries of Europe, large incorporated societies, with funds for promoting philosophical knowledge in general. Let philosophers now begin to subdivide themselves and to enter into smaller combinations. Let the several companies make small funds, and appoint a director of experiments . . . let a periodic account be published of the result of them all.'[3]

Mechanical engineering as a profession evolved out of the development of the rotative steam engine, the introduction of which in 1782 made the application of steam power possible to areas other than pumping. Prior to this date engineers had been classified as ' military engineers ', who concerned themselves with the applications of science to the arts of war, or ' civil engineers ' concerning themselves with all remaining applications of science other than military. With the development of steam power for industrial purposes the term ' civil engineer ' came to be applied solely to those studying structures of a static nature such as bridges, canals and tunnels, while the ' mechanical engineer ' was to be one who applied himself to the study and design of moving parts.

The application of steam power within industry greatly expanded in the early years of the nineteenth century, necessitating an organisation to foster the exchange of technical experience. In 1847, the year of the foundation of the Institution of Mechanical Engineers, the opinion that heat was a substance ' caloric ' was still universally held. The development of the steam engine owed little to scientific knowledge, for its evolution had been dependent on empirical trial and error methods and observation. The objects of the new institution were ' to enable mechanics and engineers in the different manufactories, railways and other establishments in the kingdom to meet and correspond, and by mutual exchange of ideas respecting improvements in the various branches of mechanical science to increase their knowledge and give an impulse to invention likely to be useful to the world '.

From the beginning the institution was interested in the organisation and retrieval of information. In addition to establishing its own library it pressed the government to improve the poor library facilities then obtaining at the Patent Office, urging it as early as 1853 ' to engage a proper staff of clerks and others to prepare abstracts and abridgments of inventions and processes '.

The institution appreciated that theory and calculation underpinned by research and experiment were needed to replace the inspired empiricism of the pioneer engineers. Accordingly, to encourage original thought in engineering science, the institution from 1875 made two annual awards of £100 and £50 for the best technical papers presented at its meetings. In the same year an annual award of £200 was made available for research into engineering science. The appointment of a research committee followed in 1879 to advise which areas of engineering should be the subjects of research projects, subcommittees being formed to direct experimental work on the hardening and tempering of steel, forming of riveted joints and friction between solid bodies at high velocity. As the institution lacked laboratory facilities, research was carried out in the workshops of firms which could be persuaded to cooperate. The last project cited was undertaken by Beauchamp Tower at the Edgware Road works of the Metropolitan Railway. Further effort was expended on alloy research, and between 1889 and 1902, when this work was transferred to the National Physical Laboratory, the institution spent £1,800 on the project, high-powered support indeed in those days.

The need for all engineers to receive a thorough grounding in the principles of engineering science was finally appreciated, and in 1912 the institution introduced an entrance examination for corporate membership.

The increasing complexity of mechanical engineering science resulting from the increasing volume of research precipitated a concentration of interests within the expanding profession, and by 1934 a number of subject groups had been inaugurated to facilitate communication between mechanical engineers working in the same fields. The present group structure comprises: applied mechanics; automatic control; education and training; hydraulic plant and machinery; industrial administration and engineering production; internal combustion engines; lubrication and wear; manipulative and handling machinery; nuclear energy; process engineering; railway engineering; steam plant; thermodynamics and fluid mechanics.

The trend towards specialisation in science and engineering is further evidenced by the proliferation of societies and institutions founded in the remaining years of the nineteenth and the early years of the present century. The Royal Aeronautical Society was founded in 1866, the Iron and Steel Institute in 1869, the Institution of Electrical Engineers in 1871, the Institute of Marine Engineers in 1889, and the Institution of Mining and Metallurgy in 1892. Professional institutions founded in the twentieth century include the Institute of Metals, 1908, the Institution of Production Engineers, 1921, and the Institute of Welding, 1923.

The learned societies and institutions are now more concerned with the education and professional well-being of their members than with the promotion of research. In this latter area their interests are more directed towards the dissemination of the results of research by the publication of scientific and technical journals documenting original work carried out by members, and the maintenance of library and information services. Many of the institutions, however, like the mechanicals, actively support research by awarding grants and fellowships. The only other societies outside the Royal Institution which maintain their own research laboratories are the Institute of Physics and the Physical Society, which recently acquired the Fulmer Research Laboratory from the British Aluminium Company, and the former Institute of Welding, which recently amalgamated with the British Welding Research Association to become the Welding Institute. The Fulmer Research Laboratory is operated on a commercial basis as a sponsored research organisation undertaking work in the fields of metallurgy, solid state physics and inorganic, physical and analytical chemistry.

The largest libraries in the fields directly relating to mechanical engineering are maintained in London by the Institution of Mechanical Engineers, the Royal Aeronautical Society, the Iron and Steel Institute, the Institution of Mining and Metallurgy, and the Institute of Marine Engineers; and in the provinces by the Institution of Engineers and Shipbuilders in Scotland and the North of England Institute of Mining and Mechanical Engineers. The remaining institution libraries are small and in some cases their use is limited to members only, although in most instances the librarians or secretaries of the bodies are willing to allow access to their collections to bona fide enquirers who are not members of the institution. Brief details of library and information services available from the institutions are included below in the classified listing. Additional details of the scope and interests of these and other scientific, technical and professional societies can be obtained from: *Scientific and learned societies of Great Britain: a handbook compiled from official sources.* Allen & Unwin, 61st edition 1964 and *Directory of British associations: interests, activities, publications . . .* CBD Research Ltd, second edition 1967/68.

AERONAUTICAL AND AEROSPACE ENGINEERING

Royal Aeronautical Society, 4 Hamilton Place, London w1. Extensive library and information services. Stock of 50,000 books and 100 current periodicals dealing with aeronautical engineering and allied subjects.

AUTOMATION AND INSTRUMENTATION

Society of Instrument Technology, 20 Peel Street, London w8.

CORROSION

British Joint Corrosion Group, 14 Belgrave Square, London sw1. Enquiry service but no library.

ENGINEERING DESIGN

Institution of Engineering Design, 28 Portland Place, London w1. Information service and small library.

FOUNDRY PRACTICE

Institute of British Foundrymen, 14 Pall Mall, London sw1. Small library of 250 books.

FUEL TECHNOLOGY

Institute of Fuel, 18 Devonshire Street, Portland Place, London w1.

HEATING AND VENTILATING ENGINEERING

Institution of Heating and Ventilating Engineers, 49 Cadogan Square, London sw1. Information service and small library of 300 books and 150 current periodicals.

IRON AND STEEL

Iron and Steel Institute, 4 Grosvenor Gardens, London sw1. Extensive library and information service. Stock of 50,000 books and pamphlets and 2,000 current periodicals. The library serves the institute and also the Institute of Metals and the Institution of Metallurgists, although the information service is limited to members of the Iron and Steel Institute. Close cooperation is maintained with the library and information services of the British Iron and Steel Research

22

Association, in that the institute is responsible for the compilation of bibliographies, the production of abstracts and translations and other literature sources, while the association's information service concentrates on matters calling for specialised technical information, the compilation of literature reviews, provision of information arising from current research and the dissemination of information from unpublished sources. The Institute has since 1966 provided an abstracts on cards service, ABTICS (Abstract and Book Title Index Card Service)[4] and from January 1969 has made available a selective dissemination of information service, ISIP (Iron and Steel Industry Profiles)[5] whereby individuals with specialised interests can receive weekly notification of specialised articles and documents which cover their field.

LOCOMOTIVE ENGINEERING

Institution of Locomotive Engineers. Merged with Institution of Mechanical Engineers in 1969 to become the Railway Division.

MARINE ENGINEERING

Institution of Engineers and Shipbuilders in Scotland, 39 Elmbank Crescent, Glasgow W2. Extensive library facilities cover all aspects of engineering in addition to the main specialisation of marine engineering and shipbuilding. Stock of over 10,000 books and pamphlets and 86 current periodicals.

Institute of Marine Engineers, Memorial Building, 76 Mark Lane, London EC3. Information service and library of 3,500 books and 140 current periodicals.

North-East Coast Institution of Engineers and Shipbuilders, Bolbec Hall, Westgate Road, Newcastle-upon-Tyne 1. Information service and library of over 4,000 books and 150 current periodicals.

Royal Institution of Naval Architects, 10 Upper Belgrave Street, London SW1. Small library for use of members.

MATERIALS HANDLING

Institute of Materials Handling, 32 Watling Street, London EC4. Information service and small library.

MECHANICAL ENGINEERING

Institution of Mechanical Engineers, 1 Birdcage Walk, London SW1. The Institution's library contains the major collection of mechanical engineering literature in the British Isles. The stock includes

80,000 bound volumes, and approximately 450 periodicals are received currently. Books, which are lent only to members of the institution and to members of the other major learned engineering societies which maintain libraries, may be borrowed by post and a subject enquiry service is also available to members. Non-members of the institution can obtain access to the library services through subscription to the group subscribers scheme, which entitles non-members working in scientific and technological disciplines related to mechanical engineering to participate in certain institution activities.

North of England Institute of Mining and Mechanical Engineering, Neville Hall, Newcastle-upon-Tyne 1. Extensive library and information services mainly covering mining technology but also extending into other fields of engineering. Stock of over 25,000 books and 10 current periodicals.

METAL FINISHING

Institute of Metal Finishing, 32 Great Ormond Street, London WC1.

METAL WORKING

Institute of Sheet Metal Engineering, John Adam House, 17/19 John Adam Street, London WC2.

METALLURGY

Institute of Metals, 17 Belgrave Square, London SW1. See Iron and Steel Institute.

Institution of Metallurgists, 4 Grosvenor Gardens, London SW1. See Iron and Steel Institute.

MINING ENGINEERING

Institution of Mining Engineers, 3 Grosvenor Crescent, London SW1.

Institution of Mining and Metallurgy, 44 Portland Place, London W1. Extensive library and information services covering mining and extraction metallurgy only. Stock of over 30,000 books and 470 current periodicals.

North of England, Institute of Mining and Mechanical Engineers, Neville Hall, Newcastle-upon-Tyne 1.

NUCLEAR ENGINEERING

British Nuclear Energy Society, 1-7 Great George Street, London SW1. No library facilities but enquiries answered by reference to the resources of the libraries of the twelve constitutional bodies, *eg* Institution of Mechanical Engineers, Iron and Steel Institute, etc.

Institution of Nuclear Engineers, 147 Victoria Street, London SW1. Small library and information service.

PHYSICS

Institute of Physics and the Physical Society, 47 Belgrave Square, London SW1. No library facilities.

PLANT ENGINEERING

Institution of Plant Engineers, 2 Grosvenor Gardens, London SW1. No library facilities.

PRODUCTION ENGINEERING

Institution of Production Engineers, 10 Chesterfield Street, London W1. Library services now administered jointly with Production Engineering Research Association, Melton Mowbray.

REFRIGERATION ENGINEERING

Institute of Refrigeration, New Bridge Street House, New Bridge Street, London EC4. Information service and small library of 1,000 books and 30 current periodicals.

TESTING AND MEASUREMENT

British Society for Strain Measurement, 281 Heaton Road, Newcastle-upon-Tyne 6. Information service for members.

Institution of Engineering Inspection, 616 Grand Buildings, Trafalgar Square, London WC2. Small library and information service.

Non-Destructive Testing Society of Great Britain, 10 Chalfont Close, Leigh-on-Sea, Essex.

Society of Environmental Engineers, 13/14 Homewell, Havant, Hampshire. Information service for members.

TEXTILE ENGINEERING

Textile Institute, 10 Blackfriars Street, Manchester 3.

WELDING

Welding Institute: see page 38.

UNIVERSITIES

The older British universities carried out a successful rearguard action against scientific education into the twentieth century. They saw themselves not as centres of research but as institutions for providing liberal education for men of a privileged class. No degree examination in science was offered at Cambridge before 1851, and until this date astronomy was the main scientific subject, with the study of heat and electricity having no place in the curriculum. Although lectures on and discussion of scientific matters subsequently took place, no official research laboratories were provided at Cambridge, professors presumably having to conduct their experiments and research in their living rooms. The Cavendish laboratory was not opened until 1874, when its moving spirit Clerk Maxwell was appointed professor of experimental physics. Financial support for the laboratory was minimal, and even in the nineteen twenties, with the advent of government assistance, expenditure on research equipment was still only apportioned in units of £50, which by 1935 had risen to units of £500. The first substantial grant to the Cavendish laboratory, enabling it to install sophisticated equipment on an engineering scale, was given by Lord Austin in 1936 and amounted to £250,000.

Developments in the redbrick universities founded after 1851 were more realistic. By 1868 Manchester had a chair of engineering, Sheffield offered an associateship in engineering from 1886, and Mason Science College at Birmingham was founded specifically ' to promote enlarged means of scientific instruction on the scale required by the necessities of the town and district '. Little research was carried out at the redbricks, but the overall position was improving and by 1873 the Devonshire Commission had stated that scientific research was an essential function of a university.

In contrast to the sluggish development of research in British universities the well-supported German technical universities provided the best opportunities for education and research in the world. German universities had by the turn of the century adopted the American principle that any problems of importance to society should be studied in a university. In 1902 the Kaiser directed the University of Charlottenburg to establish a department of machine tool design. The success of this department was immediate, and by the outset of the first world war several other such departments had been established at other German universities, giving that nation supremacy in the machine tool industry.

Although Great Britain had dominated this industry in the eighteen nineties, and pioneer work on scientific measurement was carried out at Manchester Technical College by Professor Nicholson, this initial advantage was allowed to lapse through lack of subsequent research effort. It was not until 1956 that the first department of machine tool design in a university in the British Isles was established at Manchester.

The development of German scientific and industrial research can be gauged by the volume of nineteenth century scientific publication in the German language. Most areas of science were documented by massive *Handbucher* and *Zeitschriften,* which made German the obligatory scientific language. The German ascendancy in research is further evidenced by the foundation in 1911 of the Kaiser Wilhelm *Gesellschaft* for the Promotion of Science, which administered thirty research institutions. The *Gesellschaft's* aims were the promotion of new fields of enquiry not covered by the university research laboratories, the training of graduates in research technique to equip them for research posts in industry, and to provide academic staff, who might otherwise lack the facilities and opportunities to undertake research work. Nothing comparable to the *Gesellschaft,* which was independent of state control and obtained its funds from the support of industrial organisations, bankers and businessmen, existed in Great Britain at the time. The nearest parallel was probably the cooperative research association movement, which was not to be inaugurated in this country until the following decade, by the Department of Scientific and Industrial Research.

On the eve of the first world war there were a mere 1,500 full time students studying engineering and technology in universities and colleges in Great Britain, compared with 11,000 at the German *Technisches Hochschulen.* Science and technology made little progress at British universities in the years between the two world wars. The numbers of students attending full time courses actually declined after 1928.

A measure of financial support from the government to the universities dates only from 1889, when Parliament apportioned a grant of £15,000 for the support of the recently established redbricks. Annual grants have continued from this date. The University Grants Committee, a committee of the Treasury, has existed since 1919, its members being selected on the basis of their wide experience in academic administration and industry and its terms of reference being 'to advise the government as to the application of any grants made by parliament . . . and to assist in the preparation and execution of plans for the development of the universities '. Since April 1 1965 the UGC has also handled grants for the colleges of advanced technology which were given university status following the publication of the *Robbins report*. The universities are autonomous institutions, but they receive 70 percent of their current income via the UGC in the form of government block grants.

In Great Britain the major percentage of basic research is now carried out at the universities, much of it being supported by the Science Research Council. In the physical sciences rather more than 50 percent, and in the engineering field rather less than 50 percent, of the time of university academic staff is expended on research. Examples of universities pursuing important work in the mechanical engineering field are Birmingham—machine tool problems; Bristol—strength of materials; Liverpool—applied mechanics and thermodynamics; Manchester—metal forming. A complete list of work in progress at British universities is published annually by the Department of Education and Science and the British Council, *Scientific research in British universities and colleges: volume I—physical science*. The main portion of the work lists the projects by university under broad subject groupings such as fuel technology, materials technology, mechanical engineering, and is supplemented by name indexes of research workers and a detailed subject index.

One of the pressing problems in scientific research is the relationship between the universities and industry. There is great need for each sector to reappraise the functions of the other. Sir Henry Massey has adequately summed up the present situation: ' On one hand it is said that the universities are too snobbish to take an interest in industry, and on the other that industry is so dull that nobody could be expected to want to work in it '.[6]

A recent development which will bring the universities and industry into closer contact is the establishment of industrial units at some of the universities and colleges. These units, offering initial advice to industrial firms free of charge and short term consultancy and research and development services on a repayment basis in tribology (lubrication, wear, friction, and bearing design), instrumentation and other branches of engineering, will be financed initially by the Ministry of Technology, which over the next five years will make available £1 million for their development; but it is expected that they will ultimately become self-supporting. The selected universities are Leeds—centre of tribology; University College of Swansea—industrial centre for tribology; Strathclyde—centre for industrial innovation (an effort to help small and medium sized firms to become more commercially viable); University College of North Wales, Bangor—industrial developments unit (instrumentation, materials science); and the College of Aeronautics, Cranfield—unit for precision engineering covering such subjects as the design of machine tool structures, applications of instruments and automation to machine tools and the design of high-precision engineering equipment. In addition to the above units the Ministry of Technology has established Low Cost Automation Advisory Centres at twelve universities and colleges of technology throughout the country, which aim to demonstrate the benefits of using low-cost automation control equipment.

REFERENCES

1 Hooke, Robert: *Micrographia.* 1664.
2 Sprat, Thomas: *History of the Royal Society.* 1667.
3 Priestley, Joseph: *The history and present state of electricity.* 1767.
4 Pearl, M L: *Journal of the Iron and Steel Institute, 201*(4), April 1963, p 310-316.
5 Pearl, M L: *Ibid 206*(11), April 1968, p 1151-1154.
6 Massey, Sir Henry: *Proceedings of the Royal Institution, 41*(191), p 388.

CHAPTER 3

RESEARCH IN THE UNITED KINGDOM
THE DEVELOPMENT OF GOVERNMENT SUPPORT

There is little evidence of active government support for scientific research prior to the twentieth century, effort in this sphere being limited to the work carried out in the national interest at the few state supported scientific institutions, such as the Royal Greenwich Observatory founded in 1675 for 'perfecting the art of navigation', the Geological Survey (1835) and the Meteorological Office (1845).

The first important government contribution to scientific effort was the establishment of the National Physical Laboratory in 1902 'to bring scientific knowledge to bear practically upon our everyday industrial and commercial life, to break down the barriers between theory and practice, [and] to effect a union between science and commerce'. Further evidence of increased government awareness of the vital importance of applying scientific knowledge to achieve industrial efficiency was the foundation in 1907 of the Imperial College of Science and Technology, 'to give the highest specialised instruction and to provide the fullest equipment for the most advanced training and research in various branches of science, especially in its application to industry'.

The advisory Committee for Aeronautics was appointed by the government in 1909 to advise the prime minister on the encouragement and organisation of research into the field of aeronautics; although this committee had no executive powers it was largely responsible for the establishment of an aeronautical research department at the National Physical Laboratory, and it can be cited as a precedent for the government sponsorship of key industries.

It took the shock of a world war to demonstrate to the British government that economic, and indeed national, survival would depend on close cooperation between science and industry. At the outbreak of the first world war existing British industry was not technically equipped to manufacture a whole range of essential products, from optical glass for field instruments to medicines and drugs, all of which had previously been imported from Germany, and some of which had actually been manufactured in that country from British raw materials. In 1915 the Committee for the Privy Council for Scientific and Industrial Research and an appropriate advisory council were set up in an effort to mobilise national scientific effort. The advisory council, which was mainly composed of eminent scientific representatives from those industries whose survival would now depend on scientific research, was charged with the task of apportioning government grants in support of research work.

The advisory council set up the Department of Scientific and Industrial Research (DSIR) in 1916 to implement its policy. DSIR was empowered to institute specific researches, to establish research laboratories or to encourage the expansion of the few existing research institutions, and to award research fellowships and studentships.

COOPERATIVE RESEARCH ASSOCIATIONS

The first important step to be taken by the new department was to develop a system of cooperative research associations in an effort to close the gap between science and industrial practice. The associations, formed to carry out research for the ultimate benefit of firms in specific industries, received an initial government grant, but it was envisaged that ultimately they would become self-supporting through subscriptions received from each of the industrial member firms. A grant of £1,000,000 was made to cover the initial development of the associations. This sum was exhausted by 1932, when it was agreed to continue government support in the form of an annual grant for a further period. It was not until 1943 that the British government finally resolved to form a permanent partnership with the research associations. Initially all grants were given for the general purposes of the association and were fixed according to the sum which industry was prepared to subscribe, but in 1966 a system of ' ear-marked grants ' was instituted, whereby the government contracted to support a particular research project which might yield results of considerable national benefit but which might not provide the short term benefits needed to interest private industrial development. Currently 25 percent of the combined incomes of the fifty two research associations is from the public purse.

The primary functions of the grant-aided research associations are now to enable the smaller firms to benefit from the results of research which they could not otherwise afford to carry out for themselves. To achieve this end the research associations must employ, in addition to research staff, liaison engineers capable of channelling the results of the cooperative efforts to the specialised requirements of member firms. The areas for research and the choice of priority projects are usually determined by research panels, which delegate to their sub-committees the surveillance of well-defined areas of the industry. All results of researches are made available to member firms in the form of technical reports. Members also benefit from the patents which result from cooperative research, in the form of preferential terms such as the waiving of royalties when a licensee sells the product to a member firm. The average distribution of effort throughout the fifty two grant-aided research associations is as follows:

Cooperative basic research	24 percent
Cooperative applied research	46 percent
Member and information services	23 percent
Confidential sponsored research	7 percent.

Because of the increasing volume of sponsored research which was being placed abroad by British firms immediately after the second world war, DSIR decided to encourage the grant-aided research associations to undertake this work on behalf of their members. Prior to this time doubts had been expressed as to the possibility of the associations successfully pursuing a simultaneous policy of cooperative and sponsored research. The associations now undertaking the largest programmes of sponsored work are the Production Engineering Research Associations, the British Ship Research Association, the Welding Institute and the British Iron and Steel Research Association. Other repayment work is undertaken by the associations for the National Research and Development Corporation, government departments and overseas agencies.

Member services provided by the research associations include: —

1) *Library and information services:* the majority of the associations maintain excellent libraries covering their fields of interest. The literature is exploited by the compilation of abstracting bulletins, which are sent free of charge to members who are then invited to

request on loan the full text of items relevant to their interests. The library and information section of a research association will also be responsible for the compilation and distribution of the reports bulletin, which usually gives detailed abstracts of reports which have been issued to document the results of cooperative research. Technical and production enquiries are accepted from member firms and may be answered by reference to the literature and the provision of a brief reading list or a literature survey, according to the complexity of the enquiry. Most of the research associations make the resources of their libraries available to non-members through the conventional channels of inter-library lending, and enquiries will be answered and information given to organisations who cannot reasonably be expected to become members of the association, but who are occasional users of the products manufactured or services offered by the industry which the research association serves.

2) *Liaison services:* in cases where technical enquiries cannot be adequately answered by reference to the literature, the enquirer may be visited by a liaison engineer, who from his technical experience may be able to give advice on the problem, or put the firm in touch with the most appropriate source of information. A visit from the liaison engineer could result in the firm placing a sponsored research contract with the association if no other solution to the problem was available. The Production Engineering Research Association at Melton Mowbray maintains a most comprehensive liaison service comprising of six divisions covering: heat treatment and metallurgy services; machining; metal finishing and chemistry services; metrology and inspection services; noise and vibration services; pneumatics service. The British Iron and Steel Research Association promotes two intensive consultancy services. The Steel User Section is administered from the association's Sheffield laboratories and in addition to advising members on the selection and treatment of steels, gives assistance with all problems relating to ferrous metallurgy. The Corrosion Advice Bureau is maintained at the Battersea laboratories, and advises on all aspects of the corrosion of ferrous materials.

3) *Training and education:* in many industries the grant-aided research associations are the only organisations promoting specialist training for graduates, operatives and supervisors. The British Scienti-

fic Instrument Research Association (SIRA) offers the 'Operational experience transfer' training scheme for new physics, engineering or mathematics graduates who have recently entered industry. This venture aims at giving to the young graduate industrial orientation and operational experience by working at SIRA on those real industrial problems which SIRA's information consultancy service receives from its members. The Production Engineering Research Association makes available on-the-spot advice through its production engineering advisory service. This mobile demonstration unit enables managers, designers, production staff and operatives to discuss the most advanced production practice at their own workshops at no cost to their company. Many of the associations also cooperate with universities, colleges of technology and other educational institutions to provide practical training for sandwich-course students.

The Committee of Directors of Research Associations, 24 Buckingham Gate, London SW1, was formed in 1945 to provide a forum for discussion for the directors of the associations, to act as an executive body for joint action on technical and policy matters and to effect liaison with government. The committee established an information service in 1963 to provide information to outside enquirers on the activities of the various associations. Details of the scope and services of the associations may be obtained by reference to *Technical services for industry,* an annual publication available at no charge from the Ministry of Technology.

Those associations directly serving the needs of mechanical engineers are:

AUTOMATION AND INSTRUMENTATION

British Scientific Instruments Research Association (SIRA), South Hill, Chiselhurst, Kent. Measurement and control in science and industry. Atomics and ancillaries, data handling, fluid mechanics, heat, light, magnetism, materials and design, mechanics. Publications: *SIRA review,* news on recent research for members only. *Reports* Abstracts —*SIRA abstracts and reviews,* available on subscription to non-members.

AUTOMOBILE ENGINEERING

Motor Industry Research Association, Lindley, Nuneaton, Warwickshire. Automobile engineering and related subjects such as reduction of vehicle noise, riding quality, handling and stability, aerodynamics, all aspects of vehicle safety. Publications: *MIRA bulletin*, news of current research and details of new reports. *Research reports. Foreign vehicle analysis reports.* All available to members only. Abstract bulletin: *Automobile abstracts,* available on subscription to non-members.

COAL UTILISATION

British Coal Utilisation Research Association, Randalls Road, Leatherhead, Surrey. Efficient combustion of coal, design of coal burning equipment including industrial boilers. Publications: *BCURA gazette*, details of current work, available on subscription to non-members. *Information circulars*, reports on current research, available to members only. Abstract bulletin: *BCURA monthly bulletin*, available on subscription to non-members.

CUTLERY

Cutlery and Allied Trades Research Association, Doncaster Street, Sheffield 3. Production of cutlery and small bladed instruments, scissors, and small tools with cutting edges—metallurgy, corrosion, metal plating and finishing, metal working and tool making. Publications: *Bulletin.*

FILES

File Research Council, Hoyle Street, Sheffield 3. Metallurgy and physical characteristics of files; file manufacturing processes. Publications: *Quarterly progress reports, Half-yearly bulletin.* Available to members only.

FLUID MECHANICS

British Hydromechanics Research Association, Cranfield, Bedford, Hertfordshire. Automatic control, fluid transport of solids, fundamental flow problems, hydraulic and civil engineering structures, impeller pumps and turbines, positive displacement machines and associated equipment, seals for reciprocating and rotating shafts. Publications: *Research reports, Technical notes, Progress reports, Special publications.* Available to members only. Abstracts: *Bulletin of BHMRA, Current information guides—Seals; Pumps; Fluid power; Fluidics.* All available on subscription to non-members.

FOUNDRY PRACTICE

British Cast Iron Research Association, Bordesley Hall, Alvechurch, nr Birmingham. All aspects of ironfounding, metallurgy and properties of iron castings. Publications: *BCIRA journal*, details of the association's work, available to members only. Abstracts: *Foundry abstracts*, appears in *BCIRA journal*.

British Steel Castings Research Association, Eastbank Road, Sheffield 2. Castings design and application; running, feeding and pouring techniques; solidification phenomena; patterns and pattern making; casting processes; melting and steel making; cleaning, fettling, repairing and finishing; welding in fabrication; physical and mechanical properties and testing; foundry equipment, mechanisation and layout. Publications: *Journal of BSCRA, Progress reports*, members only. *BSCRS news*, non-confidential information for non-members, *Data sheets on steel castings*, available to non-members. Abstracts: *BSCRA abstracts*, available on subscription to non-members.

HEATING AND VENTILATING ENGINEERING

Heating and Ventilating Research Association, Old Bracknell Lane, Bracknell, Berkshire. Design and performance of heating, ventilating and air conditioning systems; automatic control. Publications: *Laboratory reports, Technical notes, Information circulars*. Available to members only. Abstracts: *Thermal abstracts*, available on subscription to non-members.

IRON AND STEEL

British Iron and Steel Research Association, 24 Buckingham Gate, London SW1. Handling and preparation of iron ore; all aspects of iron and steel making including shaping, treatment and finishing, development of new steels; operation and design of plant and equipment; fuel technology, metal physics, fluid dynamics; instrumentation and automatic process control. Publications: *Research reports*, members only. *BISRA summaries*, reports of research work, available to non-members.

MACHINE TOOLS

Machine Tool Industry Research Association, Hulley Road, Hurdsfield, Macclesfield, Cheshire. All aspects of machine tool design and operation. Publications: *Machine tool research* a quarterly containing news of the design of and use of machine tools. *Notes for designers*, self-contained reviews of current practices and available equipment supported by a theoretical analysis, data is usually presented graphically for the convenience of designers. *MTIRA research reports. MTIRA library bulletin.* All publications available to members only.

MARINE ENGINEERING AND SHIPBUILDING

British Ship Research Association, Wallsend Research Station, Wallsend, Northumberland. Shipbuilding, marine engineering, ship operation and related subjects. Publications: *BSRA newsletter,* summaries of research reports and memoranda, *Technical memoranda, Research reports.* All available to members only. Abstracts: *Journal of abstracts of the British ship research association,* available on subscription to non-members.

METAL WORKING

Drop Forging Research Association, Shepherd Street, Sheffield 3. Forging techniques and related fields of machine tools, heat treatment, metallurgy, noise control, etc. Publications: *Newsletter,* information on current research, *Research reports.* Available to members only. Abstracts: *Drop forging bulletin,* available on subscription to non-members.

NON-FERROUS METALS

British Non-Ferrous Metals Research Association, Euston Street, London NW1. Fabrication and uses on non-ferrous metals including aluminium, copper, lead, magnesium, nickel, tin, zinc. Corrosion, metallography, chemical analysis, mechanical working, melting and casting, metal finishing, general metallurgy, physics and instrumentation. Publications: *BNF review,* short accounts of recently published research reports, *Research reports.* Available to members only. Abstracts: *Bulletin of abstracts,* available on subscription to non-members.

PRODUCTION ENGINEERING

Production Engineering Research Association, Melton Mowbray, Leicestershire. All methods of production, machining processes, press working, automation, work handling, assembly, inspection and measurement, plant maintenance, design and use of machine tools, cutting tools and cutting fluids, design for manufacturing, work study, application of protective coatings, cold forming, noise control. Publications: *Research reports,* members only, *PERA bulletin,* information on PERA activities and abstracts from current literature, available on subscription to non-members.

SPRINGS

Spring Research Association, Doncaster Street, Sheffield 3. Design and testing of springs; stress relaxation and fatigue of metals; materials; process and heat treatment; protective coatings. Publications: *Research reports,* members only, *Spring design and data sheets,* available to non-members. Abstracts: *Spring journal,* available to members only.

WELDING

Welding Institute, Abington Hall, Abington, Cambridge. The Welding Institute was founded in 1968 and resulted from the merging of the professional body the Institute of Welding and the former British Welding Research Association. The new organisation combines ' the staff, resources and membership of a research association devoted to technological advancement with the staff and membership of a profes- sional institute whose main interests lie in the professional knowledge, status and integrity of its members '. The institute's interests range over cooperative and sponsored research, liaison work in the transfer of research results into industrial practice, library and information services, education and training directed by the Institute's School of Welding Technology and the School of Applied Non-Destructive Testing, sponsored jointly by the institute and the Non-Destructive Testing Society of Great Britain, and the development of welding stan- dards and specifications. Welding process development, control and application; fatigue and brittle fracture; metallurgy; work measurement, standard costing and optimising techniques. Publications: *Metal con- struction and British welding journal,* a monthly technical and profes- sional journal, *Research reports,* both available on subscription to non- members. *Confidential reports,* members only.

The most successful of the grant-aided research associations serve industries composed mainly of small and medium-sized firms, there being little scope for cooperative research in those industries dominated by either two or three very large firms in direct competition, or by one massive organisation with a virtual monopoly. Some of these indus- tries are, however, served by non-grant aided research associations financed solely from the subscriptions of members, usually offering only services of a limited nature, such as testing facilities, and under- taking only limited programmes of cooperative research. The Aircraft

Research Association, for example, has available wind-tunnels for the testing of models and prototypes. There are at present fifteen of these organisations which are listed by the Ministry of Technology as being exempt from payment of income tax; those serving the mechanical engineering industries include:

Aircraft Research Association Ltd, Manton Lane, Bedford, Bedfordshire. Aerodynamic research and development using wind tunnels.

Aluminium Federation, Portland House, Stag Place, London SW1. Serves those engaged in the reduction, smelting, rolling, extrusion, drawing, casting, forging and flaking of aluminium.

British Flame Research Committee, 24 Buckingham Gate, London SW1. Study of the behaviour and properties of flames and furnaces.

Institute of British Foundrymen, Technical Research Fund, 137/139 Euston Road, London NW1. Research into the problems encountered in foundry practice. Cooperative research carried out in the works of member firms.

Permanent Magnet Association, Central Research Laboratory, 84 Brown Street, Sheffield 1. Research and development work on permanent magnet material, associated manufacturing processes and uses of permanent magnets.

Tin Research Institute, Fraser Road, Greenford, Middlesex. Research and development on all aspects of the use of tin and the improvement of existing products and the processes by which they are made.

RESEARCH STATIONS

From the time of its inception DSIR was also responsible for establishing and financing solely from public funds a number of research stations, whose terms of reference were to pursue research into areas of national interest by undertaking work necessary for the information requirements of both government departments who lacked their own research establishments and local authorities. The stations were also to undertake work on particular problems which would contribute to the efficiency of industry generally, and which may also lead to developments of national importance in specific industries.

In 1918 DSIR assumed responsibility from the Royal Society for the administration of the National Physical Laboratory, Teddington, Middlesex, now the largest of the stations, and by 1930 seven other stations had been established, whose interests ranged from road research to water pollution. The Ministry of Technology is currently responsible for maintaining seventeen research stations and establishments. The stations now all operate library and information services which will both lend specialised material and answer subject enquiries. Information concerning research work is disseminated both in the form of technical reports which are made generally available, and also by the publication of papers by station scientific and engineering personnel in the ' open literature '.

Liaison is maintained with industry by the interchange of personnel on the councils and boards of the grant-aided research associations, thus ensuring coordination of research effort by obviating the duplication of work. Copies of all reports emanating from the grant-aided research associations are received by the HQ division of the Ministry of Technology, and these are summarised in a bulletin which is distributed to all research stations and research associations.

Most of the stations, and in particular the National Physical Laboratory, cooperate with the British Standards Institution in carrying out the background research necessary in the development of national standards. The stations also undertake limited programmes of repayment work in respect of research contract work for particular sections of industry or for individual organisations and firms. Excellent examples of this cooperation between the stations and industry are for instance in the field of hydraulic machinery where the National Engineering Laboratory cooperated with a large British pump manufacturing firm to fulfil design contracts for mixed-flow pumps for the export market, and where the same station recently carried out work estimating the life of a second firm's pumps which would have otherwise involved actual life tests of several years.

The National Engineering Laboratory, East Kilbride, Glasgow, was established by the Department of Scientific and Industrial Research in 1947 to carry out basic and applied research into mechanical engineering and to provide industry with the information necessary for the solution of engineering problems. The laboratory's work includes materials at normal and high temperatures, including fatigue and creep; abrasion and wear; plasticity, including forming processes; stress analysis; fluid mechanics, including water hydraulic machinery and low speed aerodynamics; properties of fluids; heat transfer and heat exchange apparatus; machine tools, including automatic control; mechanisms; engineering metrology; bearings and lubrication; noise vibration and testing. The laboratory publishes information on its research projects in *NEL reports,* details of which are given in the periodicals *Research summaries* and *List of reports.* The annual *Heat bibliography* has been published since 1958 and is available on subscription.

The National Physical Laboratory has as its fundamental responsibility the establishment and maintenance of the nation's basic standards of measurement. Standards are determined for the basic quantities in mechanics, electromagnetics and temperature, and are derived for viscosity, hardness, pressure, etc. The laboratory's work is organised into three groups, measurement, materials, and a third group encompassing aerodynamics, autonomics, mathematics and ship divisions and a hovercraft unit. The NPL Metrology Centre in the Measurement Group, which incorporates both calibration and information services, is the focal point from which metrological services are provided to industry. In addition to the testing of measuring equipment the centre undertakes to assess the advantages and disadvantages of new designs of instruments to advise manufacturers on how to effect improvements in their products.

The *NPL quarterly* is available on request and gives information on the current work of the laboratory, in addition to giving details of current papers and reports which are available. Other important publications issued by the laboratory are the *Notes on applied science* series, the *Mathematical tables* series and the definitive series *Units and standards of measurement employed at the National Physical Laboratory.*

Other establishments concerned with aspects of mechanical engineering research maintained by the Ministry of Technology include:

Aeroplane and Armament Experiment Establishment, Boscombe Down, Salisbury, Wiltshire. Pursues work on the testing and evaluation of aircraft and aircraft equipment and systems prior to their entry into service.

National Gas Turbine Establishment, Pyestock, Farnborough, Hants. The national centre for engineering and scientific research on all matters connected with gas turbines in general and their application to aircraft propulsion in particular.

Royal Aircraft Establishment, Farnborough, Hants. The largest research centre of its type in Western Europe. Pursues and coordinates research covering the entire field of military and civil aeronautics. Provides consultative services to the aerospace industries and maintains extensive library and information services, the library stock consisting of over 25,000 books, 200,000 reports and 1,200 current periodicals. Research work is documented in *RAE technical reports*, many of which are subsequently published by HMSO as *Aeronautical Research Council papers*. *RAE news* gives information on the current work of the establishment and details of publications which are available.

SPONSORED RESEARCH INSTITUTES

In addition to the repayment work undertaken by the research associations, the research stations and to a lesser extent the universities, research work is carried out in Great Britain by independent sponsored research organisations. The existence of the research associations has worked against the development of the sponsored research organisations in the British Isles, and consequently there is little to compare with the massive organisations which exist in the United States. The chief sponsored research institutes in the United Kingdom are:

Battelle Institute Ltd, 15 Hanover Square, London W1. A subsidiary of the American institute which has its headquarters in Columbus, Ohio. Only a single agent is responsible for obtaining contracts in the United Kingdom, which are subsequently transferred to the United States. Activities cover material science and product and process development.

British Internal Combustion Engine Research Association, 111-112 Buckingham Avenue, Trading Estate, Slough, Buckinghamshire. A former grant-aided research association which became independent in 1964. Scope of work covers internal combustion engines and all general engineering subjects with particular emphasis on engine development and testing, fluid flow, instruments, stress and fatigue measurement, vibration and noise control.

Cambridge Consultants Ltd, 8 Jesus Lane, Cambridge. Contract research in all fields of engineering and the applied sciences.

Fulmer Research Institute, Stoke Poges, Buckinghamshire. Owned by the Institute of Physics and the Physical Society. The principal divisions within the institute cover physical metallurgy; foundry, metal working and ceramics; physical chemistry, thermodynamics and extraction metallurgy; physics (including electron microscopy); engineering and mechanical testing; corrosion and electrodeposition.

International Research and Development Corporation, Fossway, Newcastle-upon-Tyne 6. Covers mechanical engineering (thermodynamics, heat transfer, fluid dynamics, strain and vibrational analysis, component evaluation and development, environmental testing); materials technology; metallurgy.

Northern Research and Engineering Corporation International, Mercury House, 195 Knightsbridge, London sw7. Covers fluid mechanics, aerodynamics, combustion, heat and mass transfer, gas turbine combustion chamber design and development.

Ricardo and Co Engineers (1927) Ltd, Bridge Works, Shoreham-by-Sea, Sussex. Internal combustion engines of all types, associated work on combustion, heat flow, fuel injection etc.

Sondes Place Research Institute, Dorking, Surrey. Sponsored research in chemical and mechanical engineering, mechanical design and prototyping.

CONSULTING ENGINEERS

The assistance of consulting engineers can be enlisted for the solution of engineering problems of a practical nature. The Association of Consulting Engineers, Abbey House, 2 Victoria Street, London sw1, defines the consulting engineer as a ' person possessing the necessary qualifications to practise in one or more of the various branches of engineering who devotes himself to advising the public on engineering matters or to designing and supervising the construction of engineering works and for such purposes occupies and employs his own office and staff . . . and is not directly concerned in commercial or manufacturing interests such as would tend to influence his exercise of independent professional judgement in matters upon which he advises '. The Association publishes its *Consulting engineers' who's who and yearbook*, which gives classified listings of consultants in the various fields of engineering.

Both the Institute of Fuel, 18 Devonshire Place, London W1, and the Institute of Physics and the Physical Society, 47 Belgrave Sq, London SW1 maintain registers of members who are willing to act as consultants, and will for a fee make the necessary introductions.

RESEARCH COUNCILS

In 1918 the *Haldane report*[1] recommended that central government should extend its activities to scientific research, and in the period between the two world wars two autonomous research councils, the Medical Research Council (1920) and the Agricultural Research Council (1931), which can now be seen as the models upon which subsequently appointed research councils were designed, were established to promote research in their respective fields. The councils were composed of eminent scientists and members of the professions, their responsibility being to promote research work within the financial limitations imposed by the government, by establishing their own research laboratories and also by supporting research at the universities. The research councils are in general concerned with basic research. The organisation of these councils is based on the ' Haldane principle ', which postulates that those working on research will be more stimulated if they are coordinated within a single ministry rather than associated with the applications towards which their work is directed. The finances of research are, in addition, organised by a group of people who administer a special fund, but who have no responsibilities for the application of the results of the research—scientific research into physics is organised by a body of scientists, most of them civil servants, who are not professionally concerned with its applications. This system is diametrically opposed to the American system of ' mission-orientated research '. The Americans insist that the government department which will use the results of research should pay for it, hence the vast amount of contract research carried out inside American private industry on behalf of US government agencies.

Although the principle of government participation in the development of science and industry was now finally established, the effort being expended was not what would have been expected in view of the lessons which should have been learned in 1915. Again it took a second major conflict to revitalise national scientific effort. At the

44

beginning of the second world war a Scientific Advisory Committee to the War Cabinet was created to coordinate and intensify civil and defence research. After 1947 the defence needs of the country were directed by the Defence Research Policy Committee, while the Advisory Council for Scientific Policy was for the next eighteen years to concern itself with the needs of civilian research. Government research effort did not subside after the end of the war, and the importance now attached to this effort can be gauged from the increasing government contributions towards the cost of industrial research, and also by the expansion of the government's own scientific and industrial organisation as seen, for instance, in the foundation of the Atomic Energy Authority in 1954 and the creation of the post of Minister for Science in 1959.

A major reorganisation of British civil science was effected by the Science and Technology Act of 1965. This piece of legislation evolved from the publication of the *Robbins report*[2] on higher education and the *Trend report*[3] on the organisation of the civil service. Under the act DSIR was dissolved and its functions divided between two new departments, the Ministry of Technology and the Department of Education and Science. The Advisory Council for Scientific Policy was also dissolved and its functions taken over by the Council for Scientific Policy, which was established to advise the Secretary of State for Education and Science in the administration of his responsibilities for scientific research, and the Advisory Council on Technology, appointed to advise the Ministry of Technology. In general terms the Secretary of State for Education and Science is responsible for support of research at the universities, while the Ministry of Technology directs research of an applied nature in bringing advanced techniques and new processes to British industry.

REFERENCES

1 *Report of the Machinery of government committee.* HMSO, 1918. Cmd 9230.

2 *Committee on higher education. Report.* HMSO, 1965. Cmnd 2154.

3 *Committee of enquiry into the organisation of the civil service. Report.* HMSO, 1963. Cmnd 2171.

CHAPTER 4

RESEARCH IN THE UNITED KINGDOM
THE DEPARTMENT OF EDUCATION AND SCIENCE AND
THE MINISTRY OF TECHNOLOGY

DEPARTMENT OF EDUCATION AND SCIENCE (DES): Under the Science and Technology Act 1965 the number of research councils was increased to four by the addition of the Science Research Council and the National Environmental Research Council, while the Social Sciences Research Council was appointed in the following year. The Science Research Council was charged with ' the carrying out of scientific research, the facilitating, encouragement and support of scientific research . . . and the dissemination of knowledge in the sciences and technology '. The council's subject interests extend over the whole field of fundamental science other than those fields for which other councils are responsible. Research work is carried out in the council's own research laboratories and its terms of reference include:

i) Maintenance of these stations which include the Daresbury Nuclear Physics Laboratory, Daresbury, Cheshire and the Rutherford High Energy Research Laboratory, Chilton, Didcot, Berks.

ii) Grants to other research institutions which the council supports.

iii) Research grant to universities and colleges.

iv) Awards to post-graduate students and research fellowships.

v) UK contributions towards international organisations concerned with research.

The council maintains six research establishments which pursue research of fundamental national importance and which also provide research facilities for the universities. In addition it stimulates the application of research to industry by making grants available to finance cooperative projects between industrial firms and the universities.

DES has become the major government department concerned among other matters with the national development of libraries and informa-

46

tion services. The rapidly expanding volume of scientific and technical publication renders the task of retrieving information from the published and semi-published literature increasingly more difficult. DES appointed the Office for Scientific and Technical Information (OSTI) in 1965, its personnel being in the main transferred from those sections of DSIR engaged in information work, to provide support for tackling the scientific information problem while it was still within manageable proportions. OSTI's chief function is to promote and coordinate national documentation effort. Grants are made available for the support of research into new methods of information handling by supporting institutions engaged in feasibility tests and operational schemes; much support has already been given to the applications of computers to information retrieval. In addition to retrieval systems and selective dissemination of information (SDI) projects, which aim to keep individuals informed of current developments in their own fields, OSTI has also been active in the establishment of a series of centres to handle selected quantitative data, which lends itself admirably to storage and manipulation by computer. One such reference data centre collecting critical data on the thermodynamic values of pure compounds is maintained at Imperial College, London. Specialised information centres have also been set up in a number of fields. The high-temperature processes centre at the Houldsworth School of Applied Sciences, University of Leeds, covering the kinetics of high-temperature reactions, is one of these recently established centres whose functions are to acquire all material likely to contain information of interest to users, to arrange for expert evaluation and to classify and store material for retrieval, and finally to disseminate information according to need through current awareness publications, specialised bibliographies, answers to enquiries and state-of-the-art reports.

Other recently established information centres are the Biodeterioration Centre at the University of Aston in Birmingham, which collects and disseminates information on the deterioration of materials of economic importance by living organisms, covering such topics as bird hazards to aircraft and bacteria in lubricating oils, and the Ergonomics Information Analysis Centre at the Department of Engineering Production, University of Birmingham, which aims to meet the current needs of research and industry in ergonomics by

47

processing the current international literature on human factors in man-machine systems and physical environmental influences.

OSTI's other spheres of interest are the development of a comprehensive national library network covering science and technology, improving the presentation of scientific communication in both primary and secondary publications by, for instance, coordinating the coverage of abstracting services, the training of scientists and engineers in the use of scientific literature and the qualifications necessary for librarians and information officers. OSTI is the organ through which DES administers the National Lending Library for Science and Technology (NLL) and makes available an annual grant to ASLIB. News of OSTI activity is published quarterly in the *OSTI newsletter,* which is available without charge from the Office for Scientific and Technical Information, Elizabeth House, 39 York Road, London SE1.

ASLIB was formed in 1924, as the Association of Special Libraries and Information Bureaux, to facilitate the coordination of and systematic use of sources of knowledge and information, to promote the development of special libraries and to advise on their establishment. ASLIB now enjoys the status of a grant-aided research association. Its members include the libraries and information sections of industrial firms, and government departments, in addition to academic and public libraries. One of ASLIB's main functions is to act as a clearinghouse for technical information by putting enquirers in touch with specialist sources of information. An important feature of ASLIB's work has been the development of a number of specialised subject groups, such as the engineering, aeronautics and textiles groups, which enable information officers working in these industries to meet regularly and discuss information problems of mutual interest, and to develop the personal contacts which are so essential in the profession. ASLIB is not a professional body and does not hold examinations. Nevertheless it is vitally concerned with the training of information officers and librarians, and its annual senior information officers course provides one of the few opportunities for the scientist or engineer who has recently transferred to information work to obtain a conspectus of his newly chosen field. ASLIB maintains a library covering the fields of information work and documentation which offers a postal loan service to members. Industrial firms setting up technical libraries and information services can also request ASLIB's advice on methods of organisation and administration. One of the most recent features of ASLIB's work has been the development of a research department which conducts investigations into the problems of information handling. In collaboration with OSTI the department has selected a varied programme of research projects in the areas of publishing, storage, analysis and searching. The *sine qua non* for any project is the high probability that it will yield results which will be of immediate practical benefit. In addition to its general research programme ASLIB will also undertake, on a confidential basis, private consultancies for its members.

The National Lending Library for Science and Technology (NLL) became fully operational in 1962, and evolved out of the DSIR Lending Library Unit which was set up in 1957 to improve the availability of Soviet technical literature in the United Kingdom. NLL has recently extended its loans service into the social sciences field. The library currently acquires some 33,000 serials covering science, technology and the social sciences, and has vast holdings of technical report literature, microforms and monographs of post-graduate level. The loan service is available to industrial, academic and public libraries, but not to individual scientists at their home addresses. Every effort is made to ensure that the requester receives his literature within twenty four hours of his request being received at NLL. Since 1965 NLL has endeavoured to combat the widespread ignorance of the use of technical literature amongst scientists and technologists by promoting courses of several days duration, covering such topics as reference sources and the use of abstracting and indexing services. The most significant of these courses are those organised specifically for university science librarians. The philosophy behind the courses is to encourage university library staff to promote similar courses for undergraduates at their own universities, thus ensuring that every science graduate is given at least an introduction to the use of the literature.

The other great national library controlled by DES is the National Reference Library for Science and Invention (NRLSI), which is still not fully operational and which will not be so until a new building is erected for which a definite site has yet to be acquired. The NRLSI was instituted in April 1966 and is administered as a department of the British Museum Library. The NRLSI was based on the Patent Office Library, which was founded in 1851 and which became the largest open-access public technical library in the country. The basic NRLSI stock consisted of the holdings of the former Patent Office Library augmented by the transfer of appropriate material from the British Museum Library.

There is still an urgent need for DES to make funds available for the development of technical library resources at local level. It has already been demonstrated in an OSTI survey[1] that scientists will not travel any great distance to consult literature; thus what is needed is not simply one national reference library, but a network of regional technical reference libraries. Each regional collection should consist of the most used technical literature, which will always be readily available for consultation in the area, and comprehensive collections of abstracting and indexing services, which will enable scientists and engineers to select relevant references to the lesser used material, which can then be easily obtained on loan by the regional library from NLL. The establishment of a fully operational national reference library is however still a matter of prime urgency, in that it should coordinate and back up both the national specialist collections of the research stations and learned societies, and also the regional reference libraries, by functioning as a national referral centre and clearinghouse for technical information.

The development of the existing provincial public technical libraries, most of which operate excellent services for local industry, owes nothing to government aid. The responsibility for promoting these services, all of which, it must be remembered, serve not only industry within their boundaries but also industry situated in their environs, thus functioning as regional centres, has formerly been left to the enterprise of the local authorities prompted by the imagination of provincial city librarians. With the exception of the Sheffield Interchange Organisation (SINTO), which was founded in the nineteen thirties, the majority of the local schemes of cooperation which exist were developed throughout the nineteen fifties and sixties. In addition to providing a loan service for technical literature, subject enquiries are accepted and literature searches undertaken. Some of the schemes also issue bulletins listing recently published technical literature of potential interest to firms in the area, maintain lists of local technical translators and promote short courses on topics such as the use of technical literature. The Department of Education and Science has, however, since 1969 broadened its interests in library and information services, and, in addition to its responsibilities towards OSTI, has undertaken to promote the development of national, public and academic libraries. The new section of the department has been designated as the Library and Information Systems Branch. This development and the publication of the Dainton report[2] in June 1969, which recommends the establishment of a National Library Authority to coordinate the services and operations of the principal national library institutions, and also the development to a high standard of regional scientific information centres in all major areas based on public and academic libraries, will have a salutary effect on the development of a national information network.

Details of the services at present offered by local schemes may be found in: *Aslib directory vol I: Information sources in science, technology and commerce* edited by Brian J Wilson. Aslib, 1968. This work is the main subject guide to sources of information in the United Kingdom. A geographical listing of public libraries, academic libraries and industrial libraries and information services gives details of the subject scope of the library and details of stock. This is supplemented by a name and subject index.

THE MINISTRY OF TECHNOLOGY

The Ministry of Technology is the government department responsible for the improvement of methods of production and the expansion of productivity in the engineering industry. In pursuit of these aims its functions are to find new techniques, to evaluate research work in relation to national economic objectives, to orient the work of its own establishments to the objectives of the economy, and to disseminate the results of its work to industry. The Ministry of Technology acts as the link between government and industry by collocating the views of industry, the research associations, the research stations and the independent research organisations, to obtain as broad as possible a basis for government decisions affecting industry.

With the merging of the Ministry of Aviation into the Ministry of

Technology the department is now organised into three groups, engineering, research and aviation.

The engineering group is charged with the task of promoting the engineering industry generally. For this purpose it is organised into five industrial divisions, each responsible for a sector of the industry, and five functional divisions with pervasive interests. The industrial divisions cover: machine tools and manufacturing machinery; computers; electronics, telecommunications, instruments; vehicles and mechanical engineering production; ship-building, electrical and chemical plant. The efforts made to promote a particular industry are exemplified in the work of the machine tool division. Through this organ the ministry has given increased grants to the Machine Tool Industry Research Association and the Production Engineering Research Association in support of the development of advanced machine tools; it has made available a large grant to Birmingham University for research into advanced metal forming techniques, and has placed orders with individual firms for the development of numerically controlled machine tools. In an effort to boost the wider application of numerically controlled machine tools the division has launched a ' try-it-and-buy-it ' scheme, whereby firms are able to obtain a particular tool on a trial basis, reserving the right to return it if it proves to be unsuitable for the initial purpose of purchase. A *Register of research in machine tools* giving details of research in progress has been published and is available from room 227 at the Ministry. An Institute of Advanced Machine Tools has also been established at East Kilbride in close proximity to the National Engineering Laboratory.

The functional divisions cover, information and intelligence; standardisation (this division working in close cooperation with the British Standards Institution); economics and statistics; productivity and export services; general industrial problems (prices mergers, etc).

There are some 50,000 firms in Great Britain employing between ten and five hundred persons, and perhaps an even larger number employing less than ten. The ministry recognises their importance in terms of national productivity and is eager to see that they are given the chance to advance technologically. The small firm is however often unaware of the sources of technical knowledge available both locally and nationally. The Information and Intelligence Division of the ministry is responsible for maintaining an industrial liaison officer (ILO) service. The ILO, two-thirds of whose salary is paid by the ministry, with the remaining third being found by a local authority, is usually a previously practising scientist or engineer who has, in addition to his technical knowledge, a sound knowledge of both documentary and non-documentary sources of technical information. His function is to visit the smaller firms in his area in order to act as a trouble-shooter for production problems, and in this capacity he averages one hundred and fifty visits per year. He effects liaison between men and organisations who possess 'know-how' and those needing this knowledge. For example, a firm manufacturing boilers and heat exchangers required assistance with the calculations involved in the redesign of heat exchangers; the local ILO was able to put the firm in contact with the appropriate section of the National Engineering Laboratory, who, under contract, made a study of the specialised needs of the firm's requirements, which eventually resulted in the production costs of the redesigned heaters being reduced to fifty percent of the cost of the former products. Although the ILO may only occasionally be able to advise on a particular problem by applying his own knowledge, in many cases he is able to give immediate assistance to a firm since he is usually based on a college of technology and is thus able to enlist the help of the technical academic staff in the solution of the problem.

A recent ministry development is the *Techlink* service which aims at keeping people in specific subject areas informed of significant advances by automatically supplying them with *Techlink* sheets. The *Techlink* staff scan a vast amount of periodical and report literature and keep under surveillance the work of the research associations and stations. Important developments are reduced to their bare essentials and this information is disseminated quickly in the form of *Techlink* sheets to interested parties via the ministry's regional offices. In addition to the technical information they contain, the *Techlinks* indicate which organisation will supply additional information on the subject and they also refer to any relevant literature citations. *Techlinks* are classified into fifty two categories, such as 2-Fluid dynamics; 3-Lubrication and bearings; 4-Machinery. Participants in the scheme are simply

53

required to indicate the code numbers covering their fields of technology to the regional office.

The ministry has established a British Calibration Service at its HQ, Millbank Tower, London SEI, to provide authenticated calibration facilities for all types of measuring equipment. The service comprises a number of approved calibration laboratories in research establishments, industry and universities throughout the country who make a charge for the calibration work they undertake. The ministry HQ acts as a clearinghouse to put enquirers in touch with the laboratory maintaining suitable equipment. A similar service is the INTERLAB Cooperative Service which is administered through the regional offices. The service aims to encourage the cooperative utilisation and sharing of laboratory facilities and experience in industry. Membership, which is free, will make available to a member, within the limits of commercial security and convenience, the short term use of equipment owned by other member firms.

The Numerical Control Advisory and Demonstration Service is promoted by the ministry in cooperation with the Royal Aeronautical Establishment, Farnborough, the Production Engineering Research Association, Melton Mowbray, and Airmec-AEI Ltd, High Wycombe, in an effort to accelerate the introduction of numerically controlled machine tools into industry. The service includes assessment of a company's potential for the application of numerically controlled machine tools and demonstrations of the production of the company's own components on numerically controlled enquipment.

The Research Group is responsible for the development of the work of its research establishments. Its main areas of interest are:

i) The administration of the research stations (which function it inherited from DSIR), with the exception of the Road Research Laboratory which is administered by the Ministry of Transport, and the Radio and Space Research Station which is controlled by the Science Research Council.

ii) The promotion of the grant-aided research association movement and the administration of grants to the associations (also inherited from DSIR).

iii) Responsibility for the United Kingdom Atomic Energy Authority, a public corporation whose statutory responsibilities are divided

54

between the pursuit of research and trading activities. Under the 1965 Science and Technology Act UKAEA's terms of reference were extended to include work outside the immediate field of atomic energy. The authority now maintains twelve research establishments, the HQ of the engineering, production and reactor groups being at Risley, near Warrington, Lancs. UKAEA maintains a public information centre at 11 Charles II Street, London SW1, which operates a service covering the sources of atomic energy information. In addition to this centre, the various establishments support their own library and information services which are willing to supply unclassified information to industry. Several specialised services for industry are promoted by the authority. Free preliminary advice is given by all the specialised services, and consultancy and contract work is undertaken on repayment basis:

Aldermaston Project for the Application of Computers in Engineering, UKAEA, Blackhurst, Brimpton, Nr Reading, Berkshire.

Heat Transfer and Fluid Flow Service, Chemical Engineering Division, AERE, Harwell, Nr Didcot, Berkshire. Covers the design of boilers, condensers, heat exchange equipment and other plant.

Materials Technology Bureau, Ceramics Division, Building 35, AERE, Harwell, Nr Didcot, Berkshire. Transfers to industry the expertise acquired throughout the range of the authority's research programmes, covering electron-beam welding, hydraulic extrusion, planetary swaging, corrosion of ferrous and non-ferrous alloys, etc.

National Centre of Tribology, Reactor Engineering Laboratory, UKAEA, Risley, Nr Warrington, Lancs. Solution of friction, wear, lubrication, bearing and seal design problems.

Non-Destructive Testing Centre, AERE, Harwell, Didcot, Berkshire.

iv) Responsibility for the National Research and Development Corporation (NRDC). This is a public company, founded in 1949, consisting mainly of part time officials who are charged with the task of promoting the development and exploitation of inventions arising out of government financed research, or indeed any other invention which the corporation considers could be developed in the national interest. NRDC does not maintain its own laboratories, and the development of potentially important products and processes is effected by placing contracts with industrial firms, research associations or research

55

stations. NRDC also promotes the adoption by industry of the inventions whose development it sponsors. *Inventions for industry* is the bulletin of NRDC and includes news of the corporation's current activity and details of inventions which are available to industry under licence. Copies are available from NRDC, Kingsgate House, 66-74 Victoria Street, London SW1.

REFERENCES

1 Clements, D W G: 'Use made of public reference libraries'. *Journal of documentation*, 23(2), June 1967, p 133.

2 *Report of the national libraries committee*. HMSO, 1969. Cmnd 4028.

CHAPTER 5

TRADE AND DEVELOPMENT ASSOCIATIONS IN THE UNITED KINGDOM

Trade associations can be described as voluntary non-profitmaking organisations formed by independent manufacturing firms to protect and advance certain interests common to all member firms. The primary purpose of the trade association is to concern itself with the prosperity of the industry it serves.

The origins of trade associations in the United Kingdom can be traced to the first decade of the present century when there was a growing need for individual firms to form alliances to protect their common interests and also to cope with the problems arising from the expansion of foreign trade. The Machine Tool Trades Association was formed in 1910 to protect machine tool manufacturers from exploitation at the hands of organisers of exhibitions, the British Engineers Association, now British Mechanical Engineering Confederation, was founded in 1912 to promote the overseas interests of British engineering firms. By 1916 the Federation of British Industry had been established to act as a forum for the increasingly large numbers of trade associations and to speak for industry as a whole on important economic issues. In 1965 the Federation merged with the National Association of British Manufacturers to become the Confederation of British Industry, 21 Tothill Street, London SW1, and currently represents over four hundred individual trade associations. This body maintains an Industrial Research Committee to keep under surveillance questions relating to research as affecting industry and to recommend any action deemed necessary. The confederation has a large reference library which operates an information service covering all matters of concern to industry.

The importance of the trade associations was enhanced during the second world war when they became the channels through which the government effected liaison with particular industries. With increasing government interest in industrial matters, the importance of the trade associations has been maintained, and communication between a government sponsoring department and the industry it promotes is achieved through cooperation with the trade association.

Most trade associations offer common services to their members, including:

i) Promotion and publicity services (including the organisation of exhibitions, public relations and press services, etc).

ii) Marketing and information services—dissemination of technical information to member firms, collection of statistical information relating to the industry, production services making available the experience of one company for the benefit of other firms within the industry.

iii) Publications, *eg* buyers guides, catalogues of current plant and equipment available from member firms, newsletters including topical matter relating to the industry.

iv) Standardisation activities: many of the trade associations are the bodies responsible for standardisation within their industries and as such work in close cooperation with the British Standards Institution to develop British Standards for use in the industry. An outstanding example of this cooperation is the work of the Society of Motor Manufacturers and Traders, whose standards are often subsequently adopted by BSI as British Standards.

v) Research: although the associations do not pursue research themselves, they often have research committees whose function it is to keep the field of research under review and to work in close cooperation with the corresponding research association.

The British Mechanical Engineering Confederation (BRIMEC), 25 Victoria Street, London SW1, is the major trade association in the mechanical engineering field. BRIMEC seeks to promote the economic and commercial interests of the industry by acting as the representative national body for the industry, and also by undertaking tasks which are common to more than one of the specialist trade associations which, in addition to individual member firms, make up its membership. BMEF provides a wide range of information services and makes available to members a *Monthly bulletin* covering legal and commercial information, business oportunities and scientific and technical developments.

The Engineering Industries Association, 9 Seymour Street, London SW1, covers all the commercial aspects of engineering and assists members' firms in surmounting the financial, taxation, legal and technical problems which confront them.

Specific information on the services offered by individual mechanical engineering trade associations is given in BRIMEC's *United Kingdom trade associations: a guide, 1966*.

Development associations are non-profitmaking bodies which exist to encourage the use of materials (usually metals) and to promote their correct and efficient applications. The development associations all maintain well stocked and well administered libraries which are willing to provide literature and advice to the potential users of the materials and products which they promote. In addition, the associations all publish primary information in the form of papers, reports and conference proceedings, and secondary publications such as abstracting journals and bibliographies on specific topics. The chief development associations are the Aluminium Federation, the Cobalt Information Centre, Copper Development Association, Lead Development Association, and the Zinc Development Association.

The standard reference works covering trade, development and similar associations in the United Kingdom are: *Directory of British associations: interests, activities, publications . . . edition 2, 1967/68*. CBD Research Ltd, and Millard, Patricia: *Trade associations and professional bodies of the United Kingdom*. Pergamon Press, fourth edition 1969.

Trade and development associations whose activities either cover or impinge upon the mechanical engineering field are:

AERONAUTICAL AND AEROSPACE ENGINEERING

Society of British Aerospace Companies Ltd, 29 King Street, St James's, London SW1.

ALUMINIUM

Aluminium Federation, Portland House, Stag Place, London SW1. Comprehensive library and information services covering all aspects of the metallurgy and engineering applications of aluminium.

Association of British Aluminium & Gold Bronze Powder (Flake) Manufacturers, 6/7 New Bridge Street, London EC4.

Association of Light Alloy Refiners and Smelters, 3 Albemarle Street, London W1. Library and information service covering the metallurgy and technology of aluminium casting alloys.

AUTOMATION AND INSTRUMENTATION

British Industrial Measuring and Control Apparatus Manufacturers' Association, 22/24 Margaret Street, London W1.

Scientific Instrument Manufacturers' Association, SIMA House, 20 Peel Street, London W8.

AUTOMOBILE ENGINEERING

British Motor Trade Federation, 14 Fitzharding Street, London W1.

Society of Motor Manufacturers and Traders, Forbes House, Halkin Street London SW1.

BEARINGS

Ball and Roller Bearings Manufacturers' Association, 46 Hertford Street, London W1.

BOILERMAKING

Association of Shell Boilermakers, PO Box 498, Derby House, 12/16 Booth Street, Manchester 2.

Water-Tube Boiler Makers' Association, 8 Waterloo Place, Pall Mall, London SW1.

COBALT

Cobalt Information Centre, 7 Rolls Buildings, Fetter Lane, London EC4. Library and information service.

COMPRESSED AIR

British Compressed Air Society, 11 Ironmonger Lane, London EC2.

COPPER

Copper Development Association, 55 South Audley Street, London W1. Library and information service.

DIESEL ENGINES

Diesel Engineers and Users' Association, 18 London Street, London EC3.

FASTENERS

British Bolt, Nut, Screw and Rivet Federation, c/o Heathcote and Coleman, 69 Harborne Road, Birmingham 15.

Fasteners and Turned Parts Institute, c/o Heathcote and Coleman, 69 Harborne Road, Birmingham 15.

FOUNDRY TRADE

British Bronze and Brass Ingot Manufacturers' Association, c/o Heathcote & Coleman, 69 Harborne Road, Birmingham 15.

British Investment Casters' Technical Association, 5 East Bank Road, Sheffield 2.

Forging Ingot Makers' Association, c/o English Steel Co Ltd, River Don Works, Sheffield 4.

Foundry Trades Equipment and Suppliers' Association Ltd, John Adam House, 17/19 John Adam Street, Adelphi, London WC2.

National Brassfounding Association, 5 Greenfield Crescent, Edgbaston, Birmingham 15.

National Federation of Engineering and General Ironfounders, c/o Mann Judd & Co, 8 Frederick Place, London EC2.

Technically Controlled Castings Group, c/o John Gordon & Co, Ripley, Derbyshire.

GEARING

British Gear Manufacturers' Association, c/o Peat, Marwick, Mitchell & Co, 301 Glossop Road, Sheffield 10.

HEATING AND VENTILATING ENGINEERING

Combustion Engineering Association, 70 Jermyn Street, London SW1.

Heating, Ventilating and Air Conditioning Manufacturers' Association, Regent House, 235/241 Regent Street, London W1.

HYDRAULIC EQUIPMENT

Association of Hydraulic Equipment Manufacturers, 54 Warwick Street, London SW1. Library and information service.

British Pump Manufacturers' Association, c/o Peat, Marwick, Mitchell & Co, Glen House, Stag Place, London SW1.

Hydraulic Association of Great Britain, 160 Piccadilly, London W1.

INTERNAL COMBUSTION ENGINES

British Internal Combustion Engine Manufacturers' Association, 6 Grafton Street, London W1.

IRON AND STEEL

British Iron and Steel Federation, Steel House, Tothill Street, London SW1. Library and information service covering economic and commercial developments in British and foreign steel industries.

Crucible and Tool Steel Association, 4 Melbourne Avenue, Sheffield 10.

Joint Iron Council, 14 Pall Mall, London SW1. Information service on properties and applications of iron castings.

Steel Works Plant Association, Glen House, Stag Place, London SW1.

LEAD

Lead Development Association, 34 Berkeley Square, London W1. Library and information service.

LOCOMOTIVE ENGINEERING

Locomotive and Allied Manufacturers' Association, Locomotive House, Buckingham Gate, London SW1.

LUBRICATION

National Lubricating Oil and Grease Federation, Audrey House, 5/7 Houndsditch, London EC3.

MACHINERY

British Power Press Manufacturers' Association, Standbrook House, 2/5 Old Bond Street, London W1.

Engineering Equipment Users' Association, 20 Grosvenor Gardens, London SW1.

Food Machinery Association, 14 Suffolk Street, London SW1.

Industrial Diamond Information Bureau, Arundel House, Kirby Street, London EC1.

Paper Machinery Manufacturers' Association, 18/20 St Andrews Street, London EC4.

Textile Machinery and Accessory Manufacturers' Association, 444 Royal Exchange, Cross Street, Manchester.

MARINE ENGINEERING

Marine Engine and Equipment Manufacturers' Association, 31 Great Queen Street, London WC2.

MATERIALS HANDLING

Federation of Associations of Materials Handling Manufacturers, Glen House, Stag Place, London SW1.

Lifting Equipment Manufacturers' Association, Chamber of Commerce House, PO Box 360, 75 Harborne Street, Birmingham 15.

Mechanical Handling Engineers' Association, Glen House, Stag Place, London SW1.

METAL WORKING

Sheet Metal Industries Association, 68 Gloucester Place, London W1.

NON-FERROUS METALS

British Non-Ferrous Metals Federation, 6 Bathurst Street, London W2.

62

POWDER METALLURGY

British Metal Sinterings Association, Glen House, Stag Place, London SW1.

REFRIGERATION ENGINEERING

British Refrigeration Association, 1 Lincolns Inn Field, London WC2.

Shipowners' Refrigerated Cargo Association, Cunard House, 88 Leadenhall Street, London EC3. Library and information service.

SPRINGS

Coil Spring Federation, c/o Peat, Marwick, Mitchell & Co, 301 Glossop Road, Sheffield 10.

Spring Information Bureau, 1 Cooper Street, Manchester 2.

TOOLING AND MACHINE TOOLS

Gauge and Tool Makers' Association, Standbrook House, 2/5 Old Bond Street, London W1.

Machine Tool Trades Association Inc, 25/28 Buckingham Gate, London SW1.

National Federation of Engineers' Tool Manufacturers, Light Trades House, Melbourne Avenue, Sheffield 10.

VALVES

British Valve Manufacturers' Association, 25 Victoria Street, London SW1.

WIRE

Institute of Iron and Steel Wire Manufacturers, 10th Floor, Rodwell Tower, Piccadilly, Manchester 1.

ZINC

Zinc Development Association, 34 Berkeley Square, London W1. Library and information service.

CHAPTER 6

RESEARCH IN THE UNITED STATES
PROFESSIONAL SOCIETIES AND INSTITUTIONS

PROFESSIONAL SOCIETIES: As in Great Britain the initial interest in scientific research in the United States came from the learned societies. The first such society to be founded was the American Philosophical Society, which had its origins in a private club of scientific men founded in 1743 by Benjamin Franklin and known as the 'Junto'. The society's interests embraced every field of learning, and monthly meetings were held to discuss experiments and observations on such matters as 'new mechanical inventions for saving labour as mills and carriages, and for raising and conveying water . . . all new arts, trades and manufactures'. The American Philosophical Societies' interest in research has expanded throughout the years since its foundation, and although it does not now maintain its own laboratories it administers annual grants in support of research projects from its $175,000 Penrose research fund.

The Franklin Institute was established in 1824 as a society to stimulate the interchange of information among engineers and scientists and the interested layman, and to promote science and the 'mechanic arts' through research. The Franklin Institute founded the first important research laboratories in the United States, and since their inauguration research has been conducted in these laboratories on behalf of government and industry.

The American Association for the Advancement of Science was founded nineteen years after its British counterpart in 1848, to perform the identical function of giving a stronger and more general impulse and more systematic direction to scientific research.

In the latter half of the nineteenth century more specialised societies were set up, each concerned with a specific discipline or field of engineering. One of the first of these societies relating to mechanical engineering was the American Institute of Mining and Metallurgical Engineers, founded in 1871. The American Society of Mechanical Engineers (ASME) was established in 1880 ' to promote the art and

64

science of mechanical engineering and the allied arts and sciences and to encourage original research '. ASME research projects are currently supervised and administered in the laboratories of universities and other organisations having facilities available for cooperative research, the cost of the work being met by contributions from industry, trade associations and government agencies. The society maintains an executive research committee which supervises and coordinates the work of twenty four subject oriented committees covering such areas as lubrication or the plastic flow of metals. Each committee directs and supervises the investigation, collection, tabulation and correlation of existing data and information in its field, and when the need is sufficient it plans and directs an original research programme.

Examples of other professional and technical societies active in the field of research are the American Foundrymen's Society, established in 1896, and the American Society for Testing and Materials (ASTM), founded in 1898. The American Foundrymen's Society conducts major research on all aspects of the foundry industry at universities and in the laboratories of member firms. ASTM, whose terms of reference are the promotion of knowledge of the materials of engineering and the standardisation of specifications and methods of testing, administers a research fund with capital of $100,000, annual awards being made for outstanding original research papers.

In general, as in Great Britain, professional society interest in scientific and technical research is limited to the granting of annual awards for research papers making important contributions to the art or to supporting research fellowships. If interest in research is taken further it will, as with the example of ASME, be the recommendation and selection of specific projects to be carried out in extra-mural laboratories; few professional societies or institutions maintain their own research facilities.

Liaison in research between the professional societies in the United States is effected by the Engineers' Joint Council, 29 West 39th Street, New York 18, NY, founded in 1941. This body, whose constituent members are the twenty one national engineering societies, promotes cooperation in areas of common interest throughout the various branches of engineering. The Engineering Foundation has as its aims the advancement of research in science and engineering. This organisation, which shares the same premises as the Engineers Joint Council, was founded in 1914 and consists of a board of nineteen members representing the five major national engineering societies. The foundation initiated many of the large cooperative research councils active in American industry today, such as the Welding Research Council, 345 East 47th Street, New York, NY 10017.

In the United States the individual engineering societies do not, as a rule, maintain their own library and information services; the information needs of the twelve major engineering societies are met by the Engineering Societies Library, 345 East 47th Street, New York, NY 10017. This library, which was established in 1913 through the merging of the libraries of the American Society of Civil Engineers, the American Institute of Mining, Metallurgical and Petroleum Engineers, the American Society of Mechanical Engineers and the American Institute of Electrical Engineers, maintains the major national collection of literature in mechanical engineering. The overall subject scope of the library is defined as all branches of engineering and the related physical sciences. Each member society pays an annual sum per member for the maintenance of the collection. The stock consists of almost 200,000 bound volumes, 3,500 current technical periodicals, in addition to comprehensive sets of abstracting and indexing journals. Special collections include 600 motion picture films of research data on fluid mechanics, which are available on loan and for which a catalogue may be purchased, and a collection of over 20,000 unpublished technical papers which have been presented before sponsoring societies. Borrowing facilities are limited to members, although brief reference services are available to the general public and to industry without charge. Literature searching to the specific requirements of the enquirer is undertaken for a fee and bibliographies are prepared on request. All publications received into the library are made available for abstracting for inclusion in *Engineering index*. Joint ownership of the library is achieved by the member societies through the United Engineering Trustees, Inc, founded by the societies to administer both the library and the Engineering Foundation.

The following classified listing covers professional and technical societies with an interest in scientific and technical research and the dissemination of its results in fields related to mechanical engineering by the publication of primary journals and the support and provision of library and information services. An exhaustive listing of professional and technical societies in the United States and Canada which gives information on their research activities is: National Academy of Sciences. National Research Council: *Scientific and technical societies of the United States and Canada*. Washington, seventh edition, 1961.

AERONAUTICAL AND AEROSPACE ENGINEERING

American Helicopter Society Inc, 2 East 64th Street, New York 21, NY. Library and information service.

American Institute of Aeronautics and Astronautics, 1290 Avenue of the Americas, New York, NY 10019. Technical information service, 750 3rd Avenue, New York 7, NY.

AUTOMATION AND INSTRUMENTATION

Instrument Society of America, 313 6th Avenue, Pittsburgh, Pennsylvania. Library and information service.

AUTOMOBILE ENGINEERING

American Society of Body Engineers, 100 Farnsworth Avenue, Detroit, Michigan. Library of 10,000 volumes covering general engineering.

Society of Automotive Engineers, 485 Lexington Avenue, New York, NY. Interests cover all self-propelled ground, flight and space vehicles.

CORROSION

National Association of Corrosion Engineers, M & M Building, Houston, Texas. Library and information service.

ENGINEERING DESIGN

Industrial Designers Society of America, 60 West 55th Street, New York, NY 10019.

FOUNDRY PRACTICE

American Foundrymen's Society, Golf and Wolf Roads, Des Plaines, Illinois. Small library of 2,000 volumes and information service.

Foundry Educational Foundation, 1138 Terminal Tower, Cleveland, Ohio 44113.

HEATING AND VENTILATING ENGINEERING

American Society of Heating, Refrigerating and Air Conditioning Engineers, 345 East 47th Street, New York, NY.

INDUSTRIAL ENGINEERING

American Institute of Industrial Engineers, 145 North High Street, Columbus, Ohio.

IRON AND STEEL

Association of Iron and Steel Engineers, 1010 Empire Building, Pittsburgh, Pennsylvania.

LOCOMOTIVE ENGINEERING

American Railway Engineering Association, 59 East Van Buren Street, Chicago, Illinois.

LUBRICATION

American Society of Lubrication Engineers, 84 East Randolph Street, Chicago, Illinois. Information service.

MARINE ENGINEERING

American Society of Naval Engineers, 1012 14th Street, Washington, DC. Civil and naval membership, interests cover naval engineering including aeronautics, naval architecture, etc.

Society of Naval Architects and Marine Engineers, 74 Trinity Place, New York, NY.

MATERIALS HANDLING

International Material Management Society, 815 Superior Avenue NE, Cleveland 14, Ohio. Interests cover theory and practice of mechanical handling techniques.

Society of Packaging and Handling Engineers, 14 East Jackson Boulevard, Chicago 4, Illinois.

MECHANICAL ENGINEERING

American Society of Mechanical Engineers, 29 West 39th Street, New York, NY.

National Association of Power Engineers, 176 West Adams Street, Chicago, Illinois.

METAL FINISHING

American Electroplaters' Society, 445 Broad Street, Newark, New Jersey.

METALLURGY

American Institute of Mining, Metallurgical and Petroleum Engineers, 29 West 39th Street, New York, NY.

American Society for Metals, Metals Park, Novelty, Ohio. Small library of 4,000 volumes. The society operates a machine literature searching service covering the world's metallurgical literature, including heat treatment of metals and mechanical engineering and nuclear engineering as applied to metals. Current awareness searches are available on subscription, the search output being in the form of abstracts. Retrospective bibliographies are also available on demand.

MINING ENGINEERING

Society of Mining Engineers of AIME, 345 East 47th Street, New York, NY 10017.

NUCLEAR ENGINEERING

American Nuclear Society, 244 East Ogden Avenue, Hinsdale, Illinois 60521. Referral and consultative information services available without charge.

Atomic Industrial Forum, 3 East 54th Street, New York, NY. A technical-management association devoted to the solution of problems of the atomic industry. The forum's library contains a comprehensive collection of unclassified US Atomic Energy Commission reports which are available to the public for general use; other document services available to members only.

PHYSICS

American Institute of Physics, 335 East Ralph Street, New York, NY. A federation of the leading American societies in the field of physics, eg American Physical Society, Optical Society of America, etc. The institute maintains the Niels Bohr Library of the History of Physics, which makes on the premises reference services available to all *bona fide* enquiries. No other referral services offered.

PLANT ENGINEERING

American Institute of Plant Engineers, PO Box 185, Barrington, Illinois.

PRODUCTION ENGINEERING

American Society of Tool and Manufacturing Engineers, 20501 Ford Road, Dearborn, Michigan. Small library of 2,000 volumes and information service.

QUALITY CONTROL

American Society for Quality Control, 161 West Wisconsin Avenue, Milwaukee, Wisconsin. Library and information service.

REFRIGERATION ENGINEERING

American Society of Heating, Refrigerating and Air Conditioning Engineers, 345 East 47th Street, New York, NY.

TESTING AND MEASUREMENT

American Society for Testing and Materials, 1916 Race Street, Philadelphia, Pennsylvania.

Society for Experimental Stress Analysis, PO Box 168, Cambridge, Mass.

Society for Non-Destructive Testing, 1109 Hanman Avenue, Evanston, Illinois. Library and information service.

TEXTILE ENGINEERING

Institute of Textile Technology, Charlottesville, Virginia.

WELDING

American Welding Society, 33 West 39th Street, New York, NY. Library and information service.

CHAPTER 7

RESEARCH IN THE UNITED STATES
FEDERAL RESEARCH

In addition to the work of the professional and technical societies scientific and technical research work is currently conducted in four major areas in the United States: i) federal government laboratories; ii) private industry; iii) sponsored research and non-profitmaking institutions; iv) universities and colleges.

FEDERAL RESEARCH

Although the federal government is the major supporter of research in the United States, it carries out only a minor proportion of its research effort inside its own laboratories, transferring the bulk of this effort to the other sectors in the form of contracts. Some two thirds of *all* research in the United States is currently financed from federal sources. Prior to 1940 the three main research sectors, the federal government, private industry and the universities, tended to pursue separate research programmes, with the federal government undertaking mainly applied research, industry concerning itself with technological development, and the universities working in the field of pure science. Although these general divisions still hold true, since the second word war the lines of demarcation have become somewhat blurred, with more work of each type being undertaken in each of the three sectors.

The first US governmental grant for experimental research was made by the federal government in 1832 to the Franklin Institute of Philadelphia to finance an investigation into the causes of boiler explosions in steamboats. Continuous government interest in science in the USA can be traced to the foundation by Congress of the Smithsonian Institution in 1846. This institution was originally financed by a bequest made to the United States in 1826 in the will of James Smithson of London, to establish an organisation for the ' increase and diffusion of knowledge among men '. The institution, with the aid of an annual federal grant, has subsequently conducted scientific research in many fields, in addition to making grants to other organisations for the pursuit of research. The institution's Museum of Science and Technology includes the Division of Mechanical and Civil Engineering which has charge of technical exhibits and a study collection of machines, instruments, archives and books in these fields. Comprehensive collections of power machinery are included. The work of the division also encompasses research into early mechanical engineering literature to document engineering development.

The National Academy of Sciences was established in 1863 to advise the federal government on matters relating to scientific research; its Congressional charter specified that ' the Academy shall, whenever called upon by any Department of the Government investigate, examine and report upon any subject of science or art, the actual expense of such investigations . . . to be paid from appropriations which may be made for the purpose '. Individual governmental agencies such as the National Bureau of Standards (1901) and the National Advisory Committee for Aeronautics (1915) were later established with their own research facilities, and prior to the entry of the United States into the second world war the major proportion of federal scientific research effort was conducted inside such laboratories.

The dependence of the war effort on scientific and technical research caused the government i) to increase its expenditure on research within its own laboratories, and ii) to establish an Office of Scientific Research and Development to contract out research work to nonfederal agencies, thus utilizing the total scientific and technical resources of the nation in the war effort. The government used not only the contract method to achieve specific research objectives, but also set up completely new research centres, maintained by federal funds but administered and managed by the universities or industrial firms who were directly responsible to particular federal agencies for the work conducted in the centres they managed. Examples of those federal contract research centres carrying out important work in the mechanical engineering field are the Department of Defense's Fuels and Lubricants Research Laboratory managed by the Southwest

Research Institute, the Department of the Air Force's Arnold Engineering Development Center, managed by ARO Inc, and the National Aeronautics and Space Administration's Jet Propulsion Laboratory managed by the California Institute of Technology.

Although the Office of Scientific Research and Development was disbanded after the war, the policy of contracting work was continued by individual government departments, notably the Department of Defense, the Office of Naval Research and the Atomic Energy Commission. Extra-mural governmental sponsored research has now become the distinguishing feature of organised science and technology in the United States.

DEPARTMENT OF DEFENSE

The department was established in 1949 in succession to the National Military Establishment, with its mission as the security of the United States. Fifty percent of all federal research and development funds are expended by the department on its research, development, test and evaluation programmes, which extend over virtually the entire fields of science and technology. Most of the department's research programme is conducted or sponsored by the three military services within its jurisdiction, army, navy and air force. The responsibilities of the Department of Defense include the production, handling and dissemination of technical reports, the publishing of technical journals and the establishment and maintenance of information centres and libraries within the three main departments. The department supports several information analysis centres which 'collect, review, digest, analyse, summarise and provide advisory services . . . concerning the available scientific and technical information in well-defined, specialised fields '. These centres are distinguished from documentation centres, whose primary function is the handling of technical documents rather than the processing of the information contained in the documents. The departments information analysis centres include:

The Shock and Vibration Center, Naval Research Laboratory, Washington, DC. Collects and disseminates information on environmental factors as they affect new equipment of the government and its agencies and contractors.

Defense Metals Information Center, Battelle Memorial Institute, Columbus, Ohio. An air force sponsored centre which processes information on the properties, fabrication and applications of structural metals such as titanium, magnesium, molybdenum, in addition to analysing data on steel and steel alloys.

Mechanical Properties Data Center, Balfour Engineering Co, Sutton's Bay, Michigan. Administered by the Materials Laboratory of the Air Force Systems Command, concerned with the evaluation of data on the mechanical properties of structural metals.

Non-Destructive Testing Information Service, sponsored by the Army Materials Research Agency at Watertown Arsenal, Watertown, Massachusetts.

Enquiries on current work sponsored by the department can be addressed to the Defense Documentation Center, Defense Supply Agency, Building 5, Cameron Station, Alexandria, Virginia 22314. The primary mission of this centre is to promote the efficient interchange of military research and development information between the Department of Defense and other US government agencies. The centre maintains a collection of almost 1,000,000 technical reports, and although it does not serve the public directly, it endeavours to give the scientific and technical community access to documents on whose release there are no security restrictions.

US ARMY

In 1962 most of the major research, development and testing facilities of the US Army were grouped under a new command, the US Army Material Command, which now consists of seven subordinate commands, five of which are responsible for the research, design and testing of their own particular commodities. The Army Mobility Command has its headquarters at Warren, Michigan and is concerned with the development of aircraft, tactical and general purpose vehicles, in addition to general support equipment such as industrial engines and materials handling equipment. The Army Test and Evaluation Command at Aberdeen Proving Ground, Maryland carries out engineering design and production tests for the army and other government agencies. The Army Coating and Chemical Laboratory at

74

the Aberdeen Proving Ground is the primary army institution for research, development and evaluation of protective coatings, cleaners, fuels and lubricants for use in automobiles. The Army Materials Research Agency, Watertown, Massachusetts has as its mission the pursuit of research into metals, ceramics and armour. It adapts and improves materials already in use and develops new materials or alternate materials for strategically scarce materials. In addition to the libraries maintained by individual commands and laboratories, the Army Department maintains the Army Library, Room IA-530, Pentagon, Washington, DC 20310, which serves all Department of Defense agencies in the Pentagon. The collection of over 250,000 bound volumes, 500,000 army documents and 2,000 current technical periodicals is freely available for reference by the public.

US AIR FORCE

The United States Air Force has a growing responsibility for the support of research into all areas of science and technology related to the department's primary function of deterrence, defence and attack. Basic research is the responsibility of the Office of Aerospace Research, while applied research is the province of the Air Force Systems Command, one of the largest research and development organisations within the federal government. Enquiries concerning all aspects of air force research can be addressed to Deputy Chief of Staff, Research and Development HQ, USAF, Washington, DC 20330. Project RAND was established in 1946 to assist in the long range air force planning programmes of research and development in specific areas in the physical sciences, mathematics, economics and the social sciences. Research is carried out by the Rand Corporation, 1700 Main Street, Santa Monica, and three quarters of this organisation's activity is funded by the US Air Force. The research reports published by the corporation cover such topics as aerodynamics, engineering science, materials, nuclear engineering and physics. The Air University is concerned with the higher education of air force officers and its curricula cover a wide range of subjects including engineering and the physical sciences. An important component of the university is the Air Force Institute of Technology located at the Wright-Patterson Air Force Base, Dayton, Ohio. This institution is the chief university facility for advanced training, and laboratory facilities are maintained for research into all aspects of aeronautical and mechanical engineering. The information generated through the research programmes of the university is made available through publication and the provision of library services. The Air University Library, Maxwell Air Force Base, Alabama, gives complete university level reference service in support of the university's programmes. The stock comprises over 250,000 volumes on aerospace power, science and engineering, 500,000 reports and 3,500 current technical journals. Literature searches are made for units and individuals in the armed forces and also to others within the limits of resources available.

In addition to managing the air force basic research programme the Office of Aerospace Research promotes certain areas of applied research which have broad applications of potential value to the air force in the design and development of aerospace systems. Information generated through Office for Aerospace projects is made available through OAR publications and through library and information services. The Air Force Office of Scientific Research Technical Library, OAR Headquarters, Tempo D Building, Fourth and Independence SW, Washington, DC, is the central depository for OAR documents; library services are available to air force staff only, but an interlibrary loan service is offered. The Air Force Systems Command, Andrews Air Force Base, Washington, DC, promotes applied research programmes aimed at solving particular problems in systems development, and projects aimed at the advancement of the state of the art in the physical and life sciences and instrumentation. Library services are maintained at the command's various divisions and centres, in addition to an HQ library of reports and technical notes.

US NAVY

The US Navy pursues a continuous research, development and testing programme in order to provide its forces with the best possible equipment. More than 90 percent of its research funds are expended on applied research, and 70 percent of all projects are contracted out to industry, the universities and sponsored research organisations, with the remaining 30 percent conducted in the navy's own establishments. The Office of Naval Research, Building T3, Constitution Avenue and

17th Street NW, Washington, DC 20360, plans and coordinates research throughout the naval establishment, and conducts and supports its own research and exploratory development programmes in the physical, engineering and life sciences to design and develop training devices, training aids and their components. Enquiries concerning current research can be addressed to the Chief of Naval Research, Department of the Navy, Washington, DC 20360. The Office of Naval Research has as part of its programme the dissemination of scientific and technical information, not only within the navy but also to appropriate government departments and private organisations concerned with related research. The navy's main library is maintained at the Naval Research Laboratory, Washington, DC, the stock, which includes 85,000 bound volumes and about 300,000 technical reports, is available through interlibrary loan. The laboratory itself comprises one hundred buildings and conducts research into every area within the physical sciences.

NATIONAL AERONAUTICS AND SPACE ADMINISTRATION (NASA)

NASA, which has its headquarters at 400 Maryland Avenue, Washington, DC 20546, was established in 1958 as the successor to the National Advisory Committee for Aeronautics, to solve the problems of flight within and outside of the earth's atmosphere, and to develop, construct, test and operate aeronautical and space vehicles for research and exploration. NASA maintains several field centres and installations which are responsible for the execution and administration of research projects both through internal research and through the placing of extra-mural contracts. The Ames Research Laboratory, Moffett Field, California, directs basic and applied research in the physical and life sciences relating to the advancement of aeronautics and space technology. The Jet Propulsion Laboratory, Pasadena, California, is operated under contract by the California Institute of Technology. Enquiries covering the NASA current research effort can be addressed to Scientific and Technical Information Division, Attn Code ATSD, NASA, Washington, DC 20546. NASA headquarters and each field installation maintain libraries of varying sizes and scope to serve the needs of NASA personnel. Non-classified information is made available to the scientific community and the public through automatic distribution

77

of reports to some fifty public libraries throughout the United States. The Technological Utilization Division at NASA locates, records and analyses the scientific and engineering devices which are the by-products of the space programme and whose non-space uses are of potential value to industry. Information is disseminated to industry in the form of NASA *Tech briefs* which are distributed without charge by the Scientific and Technical Information Division to federal agencies, research and academic institutions, the technical press and other interested organisations. Important techniques are often described in detail in reports or monographs published at a later date, such as *Technical utilization report: selected welding techniques. Tools and methods developed by NASA for welding aluminum sheet and plate.* NASA SP-5003, 1963.

US ATOMIC ENERGY COMMISSION (AEC)
The AEC was established in 1946 as the successor to the US Army Manhatten Engineer District, the agency responsible for the development of atomic energy during the second world war. In addition to its defence commitments and its responsibility for the production of nuclear weapons, the AEC seeks to maximize the application of atomic energy to industrial and scientific progress by conducting and sponsoring research in nuclear science and its related disciplines, including physics, mathematics, metallurgy and materials, and by developing the peaceful uses of atomic energy. AEC research is carried out largely under contract at the universities and in private industry. The Division of Industrial Participation aims to increase private industry's participation in the peaceful development of atomic energy by informing industry of opportunities both in the private sector and in the support of government programmes, and also by acting as a central clearing-house for industry's enquiries. Each AEC installation provides an information channel between its research programme and industry's interests. The Office of Industrial Cooperation at Oak Ridge National Laboratory, Box X, Oak Ridge, Tennessee 37831, brings unclassified information to the attention of potential users. Limited assistance is also given to firms wishing to adapt new laboratory instrument designs or process techniques to general industrial use. Enquiries concerning AEC current research work can be addressed to US Atomic

Energy Commission, Division of Technical Information, Washington, DC 20545. The division operates a reference centre at Oak Ridge, Tennessee, which holds over 250,000 technical reports, and which undertakes reference services and literature searches at a fee. The AEC headquarters library at Germantown, Maryland has extensive literature collections covering nuclear science and technology, engineering and metallurgy. Interlibrary loan and limited reference services are available to non-AEC personnel.

After the second world war there was a need in the United States for a federal agency to undertake responsibility for the broader aspects of scientific research not allied to the specific interests of individual applied mission-oriented agencies, and in 1950 the National Science Foundation was established to support basic education in the sciences, and to develop and encourage the pursuit of a national policy for the promotion of basic scientific research and education. The National Science Foundation maintains no laboratories of its own, but has become the main government organ for placing research contracts with the universities. *Federal grants and contracts for unclassified research in the physical sciences*, an annual giving details of current research projects is available on request from the foundation. The foundation has important responsibilities for originating policy proposals, for financially supporting the work evolving from these proposals, and for fostering the interchange of technical information.

Government concern with technology was intensified due to the alarm created by the spectacular Russian space achievements of the late nineteen fifties, particularly the launching of Sputnik I in 1957. In this year a Presidential Science Advisory Committee, chaired by a Special Assistant to the President for Science and Technology, was set up to report directly to the President in response to specific presidential requests. The major role of the committee was that of rationalising governmental and non-governmental views on science policy, to obtain an integrated approach to problems involving science and government. On the recommendations of the first study of this committee the Federal Council for Science and Technology, a confederation of the agencies deeply involved in research and development activity, was established in 1959 to consider problems and developments in the fields of science and technology and related activities affecting more than one federal agency, or problems concerning the overall advancement of the nation's science and technology, to identify research needs and, finally, to achieve better use of facilities available nationally.

The Office of Science and Technology was founded in the Executive Office of the President in 1962 to provide a staff for both the Federal Council for Science and Technology and the Special Assistant to the President for Science and Technology. The office has complete access to all information on governmental research and development, with the overall function of evaluating and coordinating federal effort in science and technology and defining the relative roles of federal government, industry and the universities. The office is responsible for the assessment of individual scientific and technical programmes in relation to their impact on national policies. In addition to a full time staff of twelve scientists and technologists, each responsible for overseeing a particular discipline, the office calls upon the advice of governmental and non-governmental sources by liaison with the National Science Foundation, the National Academy of Sciences, the Federal Council for Science and Technology and other bodies.

The position of the National Academy of Sciences is unique. This non-governmental body has as its prime objective the advancement of science and its application for the national well-being. Its members are appointed on the basis of their outstanding achievements in science and technology. The academy speaks for the needs of science, making these known to the federal government, in addition to acting as a source of federal advice on current scientific matters. The National Research Council of the academy, which is organised into eight discipline-oriented sub-councils, concerns itself with the determination of priority areas for federal support. Like the National Science Foundation, the council maintains no laboratory facilities of its own, all research being contracted out to extra-mural laboratories. The information and data collected by the National Research Council committees during their studies, however, is often retained, and in some cases specialised information centres are established and operated on this basis.

The National Academy of Engineering was founded in 1964 under the National Academy of Science's charter to share the responsibilities of advising the federal government on engineering research.

INFORMATION SERVICES

Federal interest in the development of information services increased in accordance with the United States government's commitment to research. The National Science Foundation's concern with science information activities lies in making the results of research available to the country's scientists and engineers. The foundation's main role in this sphere is in developing effective information systems serving the scientific disciplines, presently mainly organised by the professional institutions, and in establishing workable relationships between these systems and the predominantly mission-oriented services maintained by federal agencies; thus support is currently being given to the Engineers Joint Council, Engineering Index Inc, and the United Engineering Trust, in their attempt to develop a national engineering information system.

The National Science Foundation is also vitally concerned with supporting the study of the problems associated with the storage and retrieval of information, and is currently financing several projects through its Office of Science Information Services. The development of translation programmes and the financing of cover-to-cover translations of selected Russian technical journals is yet another facet of the foundation's information activity. A useful series of thirty three pamphlets, each describing the information services operated by individual government departments, have been published by the National Science Foundation entitled, *Science information activities of Federal agencies*. Finally, financial support is given to national information services such as the Science Information Exchange and the National Referral Center for Science and Technology, in addition to maintaining close cooperation with other federal and non-federal agencies active in the field of technical information, as seen in the foundation's close liaison with the Committee on Scientific and Technical Information (COSATI) and the Committee on Scientific and Technical Communication (SATCOM).

COSATI was established in 1964 as an organ of the Federal Council for Science and Technology. Its terms of reference are to develop

81

among federal agencies coordinated but decentralised information systems. COSATI's work is effected through six panels. The information science spearhead of COSATI is the Information Science Technology Panel, whose members are the federal agencies currently operating information systems. The panel aims at the maximum exploitation of current work in information systems, and endeavours to keep abreast of all research work in this area with a view to developing new information services. The Operational Techniques and Systems Panel concerns itself with the problems of the handling and dissemination of information in the form of documents, and acts as a forum for the examination of new techniques for producing abstracting journals and indexes. The Information Analysis and Data Centers Panel is responsible for the establishment of information analysis centres and data centres within federal agencies. Data centres deal almost exclusively with the collection of quantitative data; virtually no conceptual information is processed, while the information analysis centres collect and analyse documents, and repackage the information contained in the documents in the form of digests and reviews for speedier assimilation by scientists and engineers. The National Science Foundation's Office of Science Information Service has published a guide to some of these centres: *Specialized information services in the United States; a directory of specialized information services in the physical and biological sciences.*

The Committee on Scientific and Technical Communication (SATCOM) was established in 1960 by the National Academy of Sciences and the National Academy of Engineering, with the support of the National Science Foundation, to ensure that the scientific and technical community would have a voice in the planning of national information systems. SATCOM is composed of representatives from the professional societies and institutions representing the various disciplines.

The National Referral Center for Science and Technology, Library of Congress, Reference Department, First Street and Independence Avenue SE, Washington, DC 20540, acts as an information desk for the scientific and technical community. In answer to requests the center directs the enquirer to those organisations or individuals who have the requisite technical knowledge and who are willing to share their experience. The center itself gives no technical details or bibliographical assistance; it simply acts as a clearinghouse, giving names, addresses and telephone numbers. The center welcomes registration from organisations or individuals as information sources and has, since its inception in 1962, maintained in its files details of professional societies, university and college research centres, federal and state agencies, industrial laboratories, museum specimen collections, testing stations and individual subject specialists, in addition to public libraries and document centres with special collections. Two useful directories of information resources in the United States have been published by the National Referral Center for Science and Technology: *A directory of information resources in the United States. Federal government . . . with a supplement of government sponsored information resources. 1967.* For each centre the following information is given: area of interest, library holdings, publications, information services available. The main alphabetical sequence of centres is supplemented by organisation and subject indexes. *A directory of information resources in the United States. Physical sciences, biological sciences, engineering. 1964.* An inventory of non-federal resources, each entry containing a descriptive note on the nature of the organisation, its functions and services available.

The Science and Technology Division of the Library of Congress, First Street and Independence Avenue SE, Washington, DC 20540, has a stock of over 2,000,000 volumes on all aspects of science and technology, some 20,000 current scientific and technical periodicals, 800,000 technical reports and comprehensive collections of abstracting and indexing services. The Science Reading Room is staffed by subject and language experts who offer a broad range of reference services, including instruction in the use of literature searching tools. Literature searching and the compilation of bibliographies will be undertaken on behalf of the public and industry for a prearranged fee.

The Clearinghouse for Federal Scientific and Technical Information, 5285 Port Royal Road, Springfield, Virginia 22151, is administered as a department of the Department of Commerce and exists to ensure that information generated as a result of federally supported research, and subsequently documented in technical report form, is freely available to industry, business and the general public. Comprehensive

83

collections of government sponsored research reports are held, and abstracting and indexing services giving access to the information contained in this literature are regularly issued.

The Science Information Exchange, 1730 M Street, NW, Washington, DC 20036, is financed by the National Science Foundation and administered by the Smithsonian Institution. Its mission is to assist the planning of research activity by federal and non-federal agencies, and to obviate any national duplication of research effort by collecting data relating to current research in the pre-publication stage from agencies in the United States and abroad. A brief summary consisting of a two hundred word precis, or Notice of Research Project (NRP), of each project is processed onto computer tape for subsequent retrieval. Through this service scientists and engineers in the United States can keep abreast of research effort in their subject areas. In addition to answering subject search questions in which the user will receive all NRPs related to a specific topic, the exchange offers a Periodic Mailing Service based on a ' profile ' of a requester's interest in a subject field, and involving quarterly mailing of NRPs on a continuing arrangement.

The Department of Commerce, National Bureau of Standards, Office of Standard Reference Data, Washington, DC 20234, administers the National Standard Reference Data System, a government-wide complex of activities concerned with the compilation and critical evaluation of physical and chemical data on the properties of materials and thermodynamics. This information service liaises closely with the other federally-supported data centers, such as the Thermophysical Property Data Center at Purdue University, 2595 Yeager Road, West Lafayette, Indiana 47900, which maintains a mechanised bibliographical index of world literature on the thermophysical properties of all gases, liquids and solids.

CHAPTER 8

RESEARCH IN THE UNITED STATES
PRIVATE INDUSTRY, SPONSORED RESEARCH INSTITUTIONS
AND THE UNIVERSITIES

INDUSTRIAL AND SPONSORED RESEARCH: Research is carried out in over 5,000 laboratories in industry, embracing the laboratories of individual manufacturing firms pursuing research into their own manufacturing interests, commercial laboratories specialising in research and development contract work, and non-profitmaking sponsored research organisations. The first private industrial research laboratory in the United States was founded by the General Electric Company at the turn of the century. By 1915 more than one hundred such laboratories had been established by other firms and this figure had risen to 1,600 in 1930. The National Science Foundation has published a useful directory of the commercial laboratories, *Directory of independent commercial laboratories performing research and development, 1957.* NSF 57-40. Washington, DC, US Government Printing Office, 1958. The most comprehensive current listing of industrial research laboratories of all types in the United States is *Industrial research laboratories in the United States,* edited by William W Buchanan. New York, Bowker, twelfth edition 1965. This work lists laboratories in an alphabetical sequence, giving for each entry, field of interest, major type of activity of parent company, research facilities, research personnel and laboratory staff.

The Industrial Research Council, 100 Park Avenue, New York, NY 10017, is a federation of independent industrial firms which maintain research laboratories, and exists to promote through cooperative effort improved economical and effective techniques of organising, administering and operating industrial research. The council's library is available only to member companies, but as a public service it will answer questions relating to research matters.

Some 60 percent of the research and development work carried out in industry is funded from federal sources, 90 percent of the total federal support being provided by the Department of Defense, the

Department of Health, Education and Welfare, the United States Atomic Energy Commission and the National Aeronautics and Space Administration.

The sponsored research institutes are non-profitmaking organisations whose research earnings are ploughed back into the institutions to expand the range of facilities available. The first of these organisations, the Mellon Institute, 4400 Fifth Avenue, Pittsburgh, Pennsylvania 15213, was established in 1913. Current fields of activity extend over basic research in chemistry, physics and metallurgy, in addition to environmental research for industry and government. The largest organisation in this category is now the Battelle Memorial Institute, 505 King Avenue, Columbus, Ohio 43201, established in 1925 for the ' encouragement of creative research and the making of discoveries in coal, iron, steel, zinc and allied industries '. The institute's interests now range over the entire field of science and technology. The sponsored research institute movement was stimulated by the research needs of war and from 1941 a number of these institutes were founded. Currently 50 percent of income from sponsored research contracts comes from federal sources. Other important sponsored research institutes active in the mechanical engineering field include:

Franklin Institute, Benjamin Franklin Parkway at 20th Street, Philadelphia, Pennsylvania. Applied research and engineering services in mechanical engineering, nuclear engineering and engineering materials and technology.

Midwest Research Institute, 425 Volker Boulevard, Kansas City, Missouri 64110. Pure and applied research in the physical sciences, engineering and engineering mechanics.

Southern Research Institute, 2000 9th Avenue, Birmingham 5, Alabama. Research and development in mechanical engineering, engineering physics and metallurgy.

Southwest Research Institute, 8500 Welbra Road, San Antonio, Texas. Hydrodynamics and hydroelasticity, pressure vessel technology, materials engineering.

Stanford Research Institute, Menlo Park, California. Basic and applied research in applied physics and materials science.

Details of consulting engineers practising in the United States may be obtained from the *Engineering consultants directory* published by the American Institute of Consulting Engineers, 344 East 47th Street, New York, NY 10017.

In addition to the research facilities established by individual companies and sponsored research institutes, cooperative laboratories have also been founded to serve groups of firms in a particular industry, usually organised under the auspices of a trade association. Other trade associations, lacking their own research facilities may utilise the resources of the universities on contract, or may confine their research activity to offering library and information services to disseminate to their members the results of new and relevant research work being performed nationally. The following classified listing covers trade associations with an interest in research and also cooperative research associations. Brief indication is given of library and information services where these exist. All trade associations are ready to supply statistical and commercial information on the industry which they serve. Information on the scope and activities of trade and other associations in the United States is given in *Encyclopedia of associations*. New York, Gale Research Co, fourth edition in two volumes 1964.

AERONAUTICAL AND AEROSPACE ENGINEERING

Aerospace Industries Association of America, 1725 De Sales Street NW, Washington, DC 20036. Library and information services available to members and approved enquirers.

ALUMINIUM

Aluminum Association, 420 Lexington Avenue, New York, 17, NY.

Aluminum Extruders' Council, 440 Sherwood Road, La Grange Park, Illinois 60525.

Aluminum Smelters' Research Institute, 20 North Wacker Drive, Chicago, Illinois.

AUTOMOBILE ENGINEERING

Automobile Manufacturing Association, 320 New Center Building, Detroit, Michigan 48202. Library and information service for member companies.

Motor and Equipment Manufacturers' Association, 304 West 57th Street, New York, NY 10019.

BEARINGS

Anti-Friction Bearings Manufacturers' Association, 60 East 42nd Street, New York, NY 10017.

BOILERMAKING

American Boiler Manufacturers' Association, 1180 Raymond Boulevard, Newark 2, New Jersey.

Institute of Boiler and Radiator Manufacturers, 393 7th Avenue, 10th Floor, New York, NY 10001. Library and information services for members.

COMPRESSED AIR

Compressed Air and Gas Institute, 122 East 42nd Street, New York 17, NY.

COPPER

Copper and Brass Research Association, 1271 Avenue of the Americas, New York 20, NY.

Copper Development Association, 405 Lexington Avenue, 57th Floor, New York, NY 10017. Library and information services.

DIESEL ENGINES

Diesel Engine Manufacturers' Association, 122 East 42nd Street, New York 17, NY.

FASTENERS

Industrial Fasteners Institute, 1505 East Ohio Building, Cleveland, Ohio 44114.

National Screw Machine Products Association, 2960 East 130 Street, Cleveland 20, Ohio.

Screw Research Association, 8 Mercier Road, Natick, Massachusetts.

FOUNDRY TRADE

Alloy Casting Institute, 300 Madison Avenue, New York, NY 10017. Library and information services.

American Die Casting Institute, 366 Madison Avenue, New York 17, NY.

Blast Furnace Research Inc, 2900 Grant Building, Pittsburgh, Pennsylvania.

Brass and Bronze Ingot Institute, 300 West Washington Street, Chicago 6, Illinois.

Foundry Equipment Manufacturers' Association, 1000 Vermont Avenue, Washington, DC 20005.

Gray and Ductile Iron Founders' Society, National City East 6th Building, Cleveland, Ohio 44114.

88

Investment Casting Institute, 3525 West Peterson Road, Chicago, Illinois 60645.

Malleable Founders' Society, 781 Union Commerce Building, Cleveland, Ohio.

Non-Ferrous Founders' Society, 14600 Detroit Avenue, Cleveland, Ohio 44107.

Steel Founders' Society of America, Westview Towers, 21010 Center Ridge Road, Rocky River, Cleveland, Ohio 44116. Library and information services for members.

GEARING

American Gear Manufacturers' Association, 1330 Massachusetts Avenue NW, Washington, DC 20005.

HEAT EXCHANGE EQUIPMENT

Heat Exchange Institute, 122 East 42nd Street, New York 17, NY.

HEATING AND VENTILATING ENGINEERING

Air Conditioning and Refrigeration Institute, 1815 North Fort Myer Drive, Arlington, Virginia 22209. Information service.

Industrial Heating Equipment Association, 2000 K Street, Washington 6, DC.

National Warm Air Heating and Air Conditioning Association, 640 Engineers' Building, Cleveland 14, Ohio.

Steam Heating Equipment Manufacturers' Association, 215 Central Avenue, Louisville, Kentucky 40208.

HYDRAULIC EQUIPMENT

Fluid Controls Institute, PO Box 1485, Pompano Beach, Florida.

Hydraulic Institute, 122 East 42nd Street, New York, NY 10017.

National Fluid Power Association, PO Box 49, Thiensville, Wisconsin. Library and information services.

INTERNAL COMBUSTION ENGINES

Internal Combustion Engine Institute, 201 North Wells Street, Room 1516, Chicago, Illinois 60606.

Piston Ring Manufacturers' Group, 111 West Washington Street, Chicago, Illinois.

IRON AND STEEL

American Iron and Steel Institute, 150 East 42nd Street, New York, NY. Free library, information and consultative services for members. Reference services for non-members available from the Science and

Technology Department of New York Public Library, which maintains a complete set of American Iron and Steel Institute publications for on-site reference.

LEAD

Lead Industries Association, 292 Madison Avenue, New York, NY 10017.

LUBRICATION

National Lubricating Grease Institute, 4635 Wyandotte Street, Kansas City 12, Missouri.

MACHINERY

Machinery and Allied Products Institute, 1200 18th Street NW, Washington 6, DC.

MARINE ENGINEERING

Shipbuilders' Council of America, 1730 K Street NW, Washington 6, DC.

MATERIALS HANDLING

Materials Handling Institute, Gateway Tower, Pittsburgh, Pennsylvania 15222.

MECHANICAL ENGINEERING

National Engine Use Council, 333 North Michigan Drive, Chicago 1, Illinois.

METAL FINISHING

American Hot Dip Galvanizers' Association Inc, 1000 Vermont Avenue NW, Washington, DC 20005.

Grinding Wheel Institute, 2130 Keith Building, Cleveland 15, Ohio.

Metal Treating Institute, Box 448, Rye, NY 10580.

National Association of Metal Finishers, 248 Lorraine Avenue, Upper Montclair, New Jersey 07043.

METAL WORKING

American Metal Spinners' Association, 130 Clinton Street, Brooklyn 2, New York.

American Metal Stamping Association, 3673 Lee Road, Shaker Heights 20, Ohio. Library and information services.

Forging Industry Association, 1121 Illuminating Building, Cleveland 13, Ohio.

Metal Cutting Tool Institute, 405 Lexington Avenue, New York 17, NY.

Open Die Drop Forging Institute, 440 Sherwood Road, La Grange Park, Illinois 60525.

Steel Bar Mills Association, 38 South Dearborn Street, Chicago 3, Illinois.

PIPING

Cast Iron Pipe Research Association, Prudential Plaza, Chicago, Illinois.

National Certified Pipe Welding Bureau, 666 Third Avenue, 14th Floor, New York 17, NY.

Pipe Fabrication Institute, 992 Perry Highway, Pittsburgh, Pennsylvania 15237.

Welded Steel Tube Institute, 522 Westgate Tower, Cleveland, Ohio 44116.

POWDER METALLURGY

Metal Powder Industries Federation—American Powder Metallurgy Institute, 201 East 42nd Street, New York, NY 10017. Library and information services—reference service available without charge, literature searching undertaken for a fee.

POWER TRANSMISSION

Mechanical Power Transmission Association, 3525 West Peterson Road, Chicago, Illinois 60645.

REFRIGERATION ENGINEERING

Air Conditioning and Refrigeration Institute, 1815 North Fort Myer Drive, Arlington, Virginia 22209. Information service.

TIN

Tin Research Institute, 483 West 6th Avenue, Columbus 1, Ohio.

TOOLING AND MACHINE TOOLS

Cutting Tool Manufacturers' Association, 1216 Penobscot Building, Detroit 26, Michigan.

National Machine Tool Builders' Association, 2139 Wisconsin Avenue NW, Washington 7, DC.

National Tool, Die and Precision Machinery Association, 1411 K Street NW, Washington, DC 20005.

Tool and Die Institute, 2435 North Laramie Avenue, Chicago 39, Illinois.

VALVES

Manufacturers' Standardization Society of the Valve and Fittings Industry, 420 Lexington Avenue, New York 17, NY.

Valve Manufacturers' Association, 60 East 42nd Street, New York 17, NY.

WELDING

Resistance Welder Manufacturers' Association, 1900 Arch Street, Philadelphia, Pennsylvania 19103.

WIRE

Independent Wire Drawers' Association, 1108 16th Street NW, Washington, DC.

Wire Association, 299 Main Street, Stamford, Connecticut.

ZINC

American Zinc Institute, 292 Madison Avenue, New York, NY 10017. Library and information service.

UNIVERSITY RESEARCH

The traditional responsibility for education in the United States lies with state and local government. Prior to 1941 support for university research work by federal government was limited to agricultural research, and support for the land grant colleges under the terms of the Hatch Act of 1887, which made available to each state an annual grant for the support of an agricultural experiment station and a land grant college. The land grant colleges had been established in the United States as early as 1862 by an Act of Congress, which established universities of a new type in each state, endowed with a grant of land by the federal government.

One of the basic aims of the institutions was to study any or all of the problems of society—a philosophy which was then completely alien to the universities of Great Britain. The land grant colleges turned their attention to the problems of manufacture and the mechanic arts, producing well-qualified engineers and technologists. It is significant that when in 1902 a delegation of Lancashire businessmen were writing to Andrew Carnegie, chief of the Steel Corporation of America, to report on a visit to the States, the following sentiments were expressed: ' it is not so much your enormous resources of raw material, nor the equipment in your factories, that we have come to envy. It is rather the class of educated young men who run every department in your factories. There are no such men in England '.[1] In England it was still considered somehow improper for a university to concern itself with matters mechanical. The land grant colleges were regarded with contempt in English university circles: one English academic referred to them as ' cow colleges where people learn to throw manure about and act as wet nurses to steam engines '.[2] These colleges have since grown into the great federal universities, and at their centenary half of all the PhDs and more than half of all the graduate scientists in the United States were attending their courses.

The National Science Foundation is currently the only body charged with the responsibility of furthering basic research in all fields of science, and as such is the main source of federal support for research at the universities. The mission-oriented federal agencies also place contracts with the universities for research into specific areas of science and engineering. University sponsored and other non-profit making research centres in the United States are listed in: *Research centers directory* edited by Archie M Palmer and Anthony T Kruzas. New York, Gale Research Co second edition 1965. Within an alphabetical subject arrangement the centres are listed alphabetically by name of university or institution with details of staff, areas of research and publications given. A supplement, *New research centers*, is issued regularly.

NON-FEDERAL INFORMATION SERVICES
Library and information services are administered in non-federal sectors by professional institutions, universities and colleges, public libraries and industrial firms. In addition to the two previously listed directories published by the National Referral Center for Science and Technology and the listings of professional and technical societies and trade associations operating information services, details of important collections in the mechanical engineering field can be traced by using: *Special Libraries Association: Directory of special libraries*, compiled by Isabel L Towner. 1953. An alphabetical listing of libraries with details of the collections, important subjects covered and publications issued. The main sequence is supplemented by a detailed subject index to the collections. Conditions under which the collections are available to the public are also given. And Ash, Lee (ed): *Subject collections: a guide to special book collections and subject emphasis as reported by university, college, public and special libraries in the United States,*

and Canada. Bowker, third edition 1967. Arrangement is under alphabetical subject heading, brief descriptive details of most collections are included.

More recent details of the stock, subject coverage and special collections of libraries listed in the above volumes can be obtained by consulting the biennial *American library directory* published by Bowker.

Some examples of important library and information services maintained by non-federal agencies which offer nationally available reference services covering mechanical engineering are:

Battelle Memorial Institute, 505 King Avenue, Columbus, Ohio 43201. The institute's library serves as a central source of reference for the institute's research staff, but literature searches, preparation of bibliographies, answering of reference questions, etc are also undertaken for non-members of staff on payment of a fee. The library holdings cover all aspects of mechanical and chemical engineering.

Franklin Institute, Benjamin Franklin Parkway at 20th Street, Philadelphia, Pennsylvania 19103. The institute's library holdings cover most areas of engineering, mathematics, physics and chemistry. Special collections include the Wright Brothers Aeronautical Engineering Collection. Borrowing facilities are available only to members, although the library's reference facilities are available to the public. The Library Science Information Service operates a fee-based searching, translation, abstracting service.

John Crerar Library, 35 West 33rd Street, Chicago, Illinois 60616. A library with a total bookstock of over 1,000,000 volumes and subscriptions to over 12,000 periodicals devoted exclusively to science, technology and medicine. A public reference service is offered in the Research Services Division and current scanning on specific subjects and patent prior art searches are available on an hourly cost basis through the Reference Information Service.

Massachusetts Institute of Technology, 77 Massachusetts Avenue, Cambridge, Massachusetts. The library covers all aspects of mechanical engineering, materials science and metallurgy. A reference service is offered to participants in the Membership for Industry Plan.

New York Public Library, Fifth Avenue at 42nd Street, New York, New York 10018. The holdings of the Science and Technology Division cover all branches of engineering. The library in addition maintains a complete collection of the publications of the American Iron and Steel Institute, use of which is not restricted to members of the institute. A free public reference service for brief technical enquiries is available and a fee-based service is available for more extensive enquiries and searches.

Metcut Research Machining Center, 3880 Rosslyn Drive, Cincinnati, Ohio 45209. An information centre established to correlate and disseminate machining information on metals and non-metallic materials and to provide advice on the solution of machining problems. Consulting and reference services are available at a fee.

Pacific Aerospace Library, 7660 Beverly Boulevard, Los Angeles, California 90036. A cooperative subscription library for industry sponsored by the American Institute of Aeronautics and Astronautics, which acquires and disseminates information on aerospace, aeronautical and astronautical engineering. On site reference facilities only are available to non-subscribers.

REFERENCES
1 *Proceedings of the Royal Institution, 41*(191), p 397.
2 *Ibid* p 396.

CHAPTER 9

PERIODICAL LITERATURE
DEVELOPMENT, FORMS, GUIDES

The development of the scientific periodical ran parallel to the development of the scientific society, and evolved out of the need for the scientific community to develop improved methods of communicating new advances to its members. As the numbers of scientists increased, personal communication between individuals ceased to be a practical method of keeping abreast of current developments, and an alternative channel of communication which was capable of disseminating an advance to a large audience was needed. The increasing tendency amongst scientists towards specialisation, with its attendant focusing upon more specific areas of activity, also rendered publication in the form of monograph or extended essay both impractical and uneconomic.

The first scientific periodical to be published, *Journal des scavans*, appeared in Paris on 5th January 1665. This journal included reviews of important researches and experiments, news of eminent scientists, and book reviews. It was to be the model upon which the subsequent European learned academy journals which appeared within the next decade were designed. The *Philosophical transactions of the Royal Society of London*, which commenced publication later in 1665, actually included extracts from the French journal.

The earliest scientific periodicals were interdisciplinary, containing articles and comment on all scientific matters. The first specialised journal covering physics, *Journal der physik*, commenced publication in 1790. One of the first journals to be published independently of the scientific and learned societies was the *Philosophical magazine*, a periodical which has enjoyed uninterrupted publication since its first appearance in 1798, and which is today read by engineers with a need for information on the theoretical basis of their subject.

Almost one hundred titles were appearing covering the various fields of science by the beginning of the nineteenth century. The first purely mechanical journal to be published in Great Britain was the *Mechanic's magazine*, which was founded in 1823 to give mechanics ' a better acquaintance with the history and principles of the arts they practise '. Issues of the journal were to include ' accounts of new discoveries, inventions and improvements, with illustrated drawings, explanations of secret processes, economical receipts, practical applications of mineralogy and chemistry, plans and suggestions for the abridgment of labour, reports on the state of the arts in this and other countries, memoirs and occasionally portraits of eminent mechanics, etc '. The new journal rather patronisingly announced that ' communications from intelligent mechanics will be thankfully received '. The *Mechanic's magazine* was an offshoot of the mechanics institute movement, and as such its contents were directed at the artisan classes; nevertheless this journal can be cited as the precursor of today's commercially published technical and trade journals.

The first society journal dealing specifically with mechanical engineering to be published in Great Britain was the *General proceedings of the Institution of Mechanical Engineers*, which first appeared in 1847. The initial volume printed the text of papers which had been presented before the institution during the period from 1847 to 1849, the subjects of the papers including a description of an hydraulic starting apparatus, the balancing of locomotive wheels, the prevention of explosions in boilers and the fallacies of the rotary engine.

By the middle of the nineteenth century it had been estimated that almost one thousand scientific and technical journals were being published throughout the world. The *Engineer*, which has for over one hundred years been one of the practitioner's most valuable sources of current information, was founded in 1856, not ' to furnish a dry register of the progress of machinery ', but ' to represent effectively the industrial activity in which we live, to keep pace with the progress of improvements and developments in all the departments of the arts and manufactures which contribute to our material comfort '. The *Engineer* was a journal for the professional man rather than for the artisan, and its first editorial considering the interdependence of science and technology could well be used today to illustrate the importance to the mechanical engineer of *all* the literature of science and technology. ' The professional engineer finds indeed that the data with which he works are to be gathered alike from the most homely precepts of everyday experience, and the remotest provinces of the physical sciences. He is necessarily a mechanic, but not a mechanic merely; he has need of the deductions of chemistry as well as—as far at least as they are subservient to his precise knowledge—the material properties and agencies which it is his business to apply and direct.' Important features of the early issues of the *Engineer* were the ' systematic expositions of particular arts and branches of manufactures showing their state of development ', and the patent section which gave detailed information on recently published specifications.

The other important current weekly journal on general engineering topics, *Engineering*, first appeared in 1866. *Nature*, the most influential of all the journals covering advances in the pure sciences was first published in 1869. Its editorial policy, as dictated in the first issue, has remained constant since that time: ' to aid scientific men by giving early information of all advances made in any branch of natural knowledge throughout the world, and by affording them an opportunity of discussing the various scientific questions which arise. . . .'.

In the United States the *American mechanic's magazine*, which was a frank imitation of the London *Mechanic's magazine*, was founded in 1825 as ' a digest of mechanics and scientific progress '. This journal subsequently became the *Journal of the Franklin Institute*, a publication which was of particular significance in the early years of the nineteenth century because of its descriptions of newly published American patent specifications. Prior to 1843 the United States Patent Office omitted details of the claims of patent grants from its official publications, which included only details of title and the name of the patentee. The other important early mechanical engineering journal to be published in the United States, the *Scientific American*, commenced publication in 1845 with its subtitle as ' the advocate of industry and journal of mechanical and other improvements '. In its early years the *Scientific American* was published as both a journal of opinion, whose mission was the debunking of perpetual motion cranks and other quacks, and as an organ for publicising new mechanical inventions. The mechanical bias has long since disappeared and for many years the journal has concentrated on publicising the importance of science to the community in general, consistently maintaining impeccable standards of content and presentation. The first American society publication to be devoted exclusively to mechanical engineering, the *Transactions of the American Society of Mechanical Engineers,* first appeared in 1880.

Current technical periodicals can be conveniently divided into three groups: i) periodicals published by learned societies and institutions, ii) independent periodicals published by commercial publishing firms, iii) house journals published by industrial firms and organisations.

LEARNED SOCIETY JOURNALS

Periodicals in this group publish papers presented before meetings of the parent body in a publication usually termed ' journal ' or ' transactions ' or ' proceedings '. Strict editorial control is effected over the content of these publications by an editorial board composed of eminent practising members of the profession, to ensure that any material accepted for publication represents an original and worthwhile contribution to the art. The learned societies see themselves as the guardians of the standards of publication within their fields, and it is in their journals that the most important contributions to the literature of science and technology appear. In two citation examination projects carried out in the field of mechanical engineering literature, in which an analysis was made of the articles and documents cited by authors of papers published over a period of ten years in i) the *Proceedings of the Institution of Mechanical Engineers,*[1] and ii) the *Transactions of the American Society of Mechanical Engineers,*[2] in both instances the six periodicals most frequently cited were published by the American Society of Mechanical Engineers, the Institution of Mechanical Engineers and the Royal Society of London.

Some societies include information on the day-to-day activities of the parent body in the same journal which publishes original papers, while others issue a separate ' general purpose journal ' to act as the link between the society and its members. An excellent example of such a journal is the Institution of Mechanical Engineers' monthly *Chartered mechanical engineer,* each issue of which contains regular features such as notes of institution meetings, news of activities of the groups, articles of general interest in the form of reviews of progress (*eg* computer applications in mechanical engineering), summaries and synopses of papers to be presented at future meetings, overseas developments, notes on personalities within the profession, correspondence, accessions to the library and library notes. The general purpose journal of the American Society of Mechanical Engineers is *Mechanical engineering* which, in addition to society news, includes news briefs on engineering developments culled from current journals and reports, features which are likely to be of general interest, and abstracts of technical papers presented at society meetings. All papers presented before meetings of the American Society of Mechanical Engineers are originally issued as preprints. These papers are divided into two groups—permanent interest material which is ultimately published in the *Transactions*, and current interest material which is available in preprint form from the society for a period of up to ten months after publication. All preprints are initially announced in the January issue of *Mechanical engineering*, and some papers from both groups appear in this journal. Almost one half of all papers published are of the current interest type.

COMMERCIALLY PUBLISHED JOURNALS
Most fields of technology are served by at least one commercially published journal. These periodicals, often referred to as trade journals, are published for profit, although only in rare cases will the total numbers of subscriptions paid by readers cover the costs of production. The publisher's profit is invariably drawn from advertising revenue, and advertisements are a regular feature of all commercially published technical journals. It has been pointed out that mechanical engineers greatly value advertisements appearing in periodicals as a method of keeping informed of new products and processes appearing on the market.[3] This class of periodicals also strongly features articles on topics of current interest to the industry they serve. The articles are usually written by staff technical journalists, and tend to be broader in interest than the papers which are published in the primary journals; one such article will often summarise in an easily digestible form the content of several more specific but related papers which have originally appeared in the primary journals. Other regular features in the typical commercially published journal will be information on new plant and equipment,

descriptions of recent national and foreign patent specifications of relevance to the industry, notifications of new British and foreign standards, prices of raw materials, selections from recently issued trade literature, abstracts of important articles from other periodicals, book reviews and news of personalities within the industry.

In addition to the trade journals a small number of specialist firms, Academic Press, Pergamon Press, Butterworth, Elsevier, etc, publish archival journals commercially, usually including papers in several languages. The content of these journals is invariably vetted by a board of editors drawn from the scientific community, in emulation of the learned society method of effecting content control. Excellent examples of such publications are *Journal of fluid mechanics* (Cambridge University Press), *International journal of machine tool design and research* (Pergamon Press) and *Wear* (Elsevier). Some of these publishers also issue journals on behalf of a society, charging to the society the cost of individual members subscriptions and collecting for themselves the revenue from non-member subscriptions.

The controlled-circulation journal is a particular type of commercially published periodical issued by a publisher to promote the products of several manufacturing firms engaged in a particular industry. These journals will be mailed regularly, free of charge, to senior technical personnel employed by those companies who are the potential purchasers of the products advertised in the pages of the journal. Controlled-circulation journals consist almost entirely of information on new products and advertisements, their readers are invited to fill in pre-paid postcards to obtain further information on particular items. Any additional information contained in these journals will consist of very general articles summarising trends within the industry or news of exhibitions and trade fairs.

HOUSE JOURNALS

The house journal is a publication issued by an industrial company or organisation to project and enhance the image of the organisation.

House journals are made available, usually on a subscription free basis, to the potential customers of the company or to the potential users of the services the organisation purveys. One type of house journal is the 'prestige journal' which seeks to promote the interests of the parent body, not by directly publicising its products, but more subtly by its association with standards of excellence both in presentation and content. Impressive presentation is achieved by adopting the highest standards of typography and illustration, and excellence of content either through releasing accounts of advanced work carried out by company technical personnel, or by presenting review articles by highly qualified specialists on various aspects of science and technology not necessarily connected with the company's manufacturing programme. Excellent examples of prestige journals are the admirably designed *Endeavour* published by Imperial Chemical Industries Ltd in five separate language editions as a review of progress in science, the *Journal of science and technology*, published by the General Electric and English Electric Co and including specialised signed articles on various aspects of technology, and the *Ball bearing journal* published by the Skefko Ball Bearing Co as a quarterly review of ball-bearing engineering.

A more direct approach to the promotion of the interests of the organisation is achieved through the journal publishing information on the uses and applications of the company's products. Examples of this type of journal are the *Hawker-Siddeley technical review, BSA tools journal* and *Dexion news*. The categories of house journal referred to above are known as ' external ' house journals in that they are designed for circulation to non-company personnel. The ' internal ' house journal is of little value as a source of technical information, as this is produced for company personnel with the express purpose of promoting a sense of community within the organisation by acting as the link between management and employees and by publishing features on company group and social activities. House journals are a useful source of technical information. In addition to being usually available at no cost, they give information on the products and activities of a company's rivals, and in many cases publish articles of real technical value. A selection of useful house journals is given in the following pages. It is estimated that some 2,500 house journals are published currently within the United Kingdom and approximately 10,000 in the United States. The standard listings of house journal are:

British Association of Industrial Editors yearbook. The Association, 12 Thayer Street, London W1. An alphabetical listing of firms publishing house journals, supplemented by a title index, but unfortunately without either subject or industry indexes. (A useful directory of British house journals compiled by Isabel J Haberer and including much descriptive detail appears in *Progress in library science 1967*, edited by R L Collison. London, Butterworth, 1967. pp 17-96.

Gebbie house magazine directory, 1965-1966-. House Magazine Publishing Co Inc, Sioux City, Iowa, USA. Published every two years, an alphabetical listing of American and Canadian firms publishing edited by R L Collison. London, Butterworth, 1967. p. 17-96.

Periodicals play a key role in the pattern of primary publication as they enable information to be disseminated speedily and conveniently: speedily in that news of recent developments can be published with a delay limited to days rather than months as with the monograph, and conveniently in that most accounts of scientific and engineering progress can be organised into an article of four or five thousand words, the average length of a paper in the primary journals. The archival or the primary journals are mainly published by the scientific and engineering societies, and these publications document the important advances in science and technology, while the commercially published journals are the news organs publishing topical information of an immediately practical nature. To the specialist the periodical is the tool which reproduces another's thought and experience: it can enable the engineer to reproduce the other man's work or at least to build on his experience. It has been recorded that George Westinghouse resolved to develop a better type of locomotive brake after witnessing a railroad accident. He subsequently read an article in a technical journal concerning the use of compressed air in tunnelling through a mountain. The article gave him the idea which led him to the development of the air brake. Henry Ford obtained the initial idea for an improved engine for his automobile after reading a detailed technical article describing the ' Kane-Pennington motor ', then a revolutionary type of internal combustion engine.

Periodicals supplement the monograph as a form of publication in that it would not always be economically possible to publish short accounts of research or surveys of progress in highly specific fields as separate publications. By regularly scanning the important news and primary journals covering his field the mechanical engineer can ensure that he is keeping abreast of new developments relevant to his interests. He can in addition keep himself aware of developments in other fields of science and engineering which might have possible applications in his own field by regularly scanning the more general periodicals covering science and technology. Periodicals are also important retrospectively, for although a high percentage of information originally appearing in the periodical literature is ultimately summarised in a monograph covering the subject field in general and published maybe some years after the original item, much material never reappears in book form. Sets of bound volumes of periodicals must, therefore, be maintained in libraries to cater for the retrospective demand for information, and to permit librarians to undertake state of the art searches of the literature.

The most comprehensive listings of periodicals are the national union catalogues, which record the holdings of important national, municipal, university and special libraries. These catalogues can be used both to ascertain locations for sets of a specific title, and also to check bibliographical details such as changes in title or verification of the place of publication. They are of little use in subject selection as they are invariably simple alphabetical listings of titles, merely giving for each title details of the holdings of cooperating libraries. The *British union catalogue of periodicals* (BUCOP), London, Butterworth, four volumes 1954 is the major key to the holdings of libraries in the British Isles. A supplement to the work was published in 1960 listing titles appearing after publication of the basic work, emendations to the basic work and some titles not previously reported. The five volumes list some 140,000 titles held in 440 cooperating libraries. Since 1964 BUCOP has published a quarterly supplement listing new periodical titles and titles published after 1960; this cumulates annually and quinquennially, and a separate annual listing of titles in science and technology is also issued as the *World list of scientific periodicals*.

The holdings of American and Canadian libraries are listed in *Union list of serials in libraries of the United States and Canada*, New York, H W Wilson, third edition in five volumes 1965. No further editions of this work, which lists some 150,000 titles published up to 1949, are planned, but it has been continued as *New serial titles*, a monthly list, prepared and published by the Library of Congress, which commenced publication in 1950 and which has annual, quinquennial and decennial cumulations. *New serial titles* is an alphabetical listing of titles held in the Library of Congress and other cooperating libraries in the United States. A separate monthly edition with a subject breakdown of titles arranged by the Dewey decimal classification is also available.

Two other important catalogues of periodicals in science and technology are:

National Lending Library for Science and Technology: *Current serials received by NLL.* March 1967. London, HMSO, 1967. An alphabetical title listing without holdings; and

National Federation of Science Abstracting and Indexing Services: *A list of serials covered by members of the NFSAIS.* Two volumes 1962. An alphabetical listing of titles covered by member services with an indication for each title of the covering services.

NATIONAL AND SUBJECT LISTINGS

The above tools are of limited use in the subject selection of periodicals. Of more value in this context are the national lists of periodicals published in a particular country, which invariably incorporate subject groupings in their arrangement.

National listings: Great Britain. Willing's press guide, London, James Willing Ltd, is an annual directory with quarterly supplements giving publication details of periodicals, newspapers and annuals published in the British Isles. The arrangement is alphabetical by title with a subject index under broad headings. Select listings of Commonwealth and foreign titles are also given, these being listed under country only. A guide giving more detailed information on individual titles is Toase, Mary (ed): *Guide to current British periodicals,* London, Library Association, 1962. This publication is eminently suitable for subject selection as it is arranged by the Dewey decimal classification and includes information on the average number of pages per volume, scope, coverage by abstracting and indexing services, availability of indexes and supplements, etc. A useful checklist of British scientific and technical periodicals arranged under broad subjects is provided as an appendix to Ball, I D L (ed): *Industrial research in Great Britain,* London, Harrap Research Publications sixth edition 1968. New British technical periodicals can be identified each week by checking under the form heading ' periodicals ' in the index to the *British national bibliography.*

United States: Standard periodical directory, New York, Oxbridge Publishing Co, is an annual directory listing over 20,000 periodicals published throughout the United States and Canada under some 200 alphabetically arranged subject groups. Information given for each title includes scope and circulation, and house journals are included as a separate listing. The main sequence is supplemented by a title index. More specific information is given for 320 titles, most of which are published in the United States in—Martin, Ralph C and Jett, Wayne: *Guide to scientific and technical periodicals: a selected and annotated list of those published in English,* New York, Swallow, 1963. Although the information given for each title on editorial policy, length of average article, types of illustration etc is primarily intended for the potential technical author or the advertiser, this work is of obvious relevance in the subject selection of periodicals; arrangement is within six broad subject groups, which are again further subdivided and a title index is included.

France: *Annuaire de la presse française et étrangère et du monde,* Paris, Annuaire de la Presse. Classified arrangement of French periodicals and highly selective listings of titles from other countries. *La presse française 1965: guide général méthodique et alphabétique,* Paris, Hachette is an alphabetical and classified listing of French titles.

Germany: *Deutsche presse 1961: zeitungen und zeitschriften,* Berlin, Duncker & Humbolt. A broad subject grouping of 6,500 West and East German periodicals supplemented by a title index. German scientific, technical and medical periodicals and newspapers are listed annually in Saarbach, WE GmbH: *Subscription catalogue,* Berlin, an alphabetical title arrangement with subject index.

USSR: *Gazety i zhurnaly SSSR* (Newspapers and magazines of the USSR) an annual classified listing of Russian current periodicals available for purchase through the distributors Collets Holdings. Titles are given in Russian, in transliteration and in English, with title indexes in Russian, English, French, Spanish and German.

The most comprehensive subject guide to periodicals published throughout the world is *Ulrich's international periodicals directory: a classified guide to periodicals, foreign and domestic,* New York, Bowker, 2 vols, thirteenth edition 1969-1970, which gives detailed information (including circulation figures, coverage by abstracting services, presence of illustrations and advertisements, inclusion of bibliographies, etc) for some 40,000 titles which are arranged alphabetically by title within broad subject groups. A title and subject index supplements the main sequence. Volume I of the *International media guide: the world wide media service,* New York, R A Hill Co Ltd, lists some 5,000 technical, professional and trade periodicals with international and national circulation. The international and national sections are each divided into 96 subject groupings with further subdivision of the national section by county. Each annual issue of *Engineering index* contains alphabetical listings of titles covering engineering in general, while titles of periodicals on mechanical engineering appear in each January issue of the *Chartered mechanical engineer* (title listing subdivided by country).

REFERENCES

1 Investigation undertaken by the author at Liverpool Polytechnic, Department of Library and Information Studies.

2 Burton, R E: 'Citations in American engineering journals, II. Mechanical engineering'. *American documentation, 10,* 1959, p 135-137.

CHAPTER 10

SCIENCE: The two most useful weekly periodicals for keeping abreast of developments in the sciences are the British *Nature* (Macmillan (Journals) Ltd, Little Essex Street, London WC2), and the American *Science* (American Association for the Advancement of Science, 1515 Massachusetts Avenue NW, Washington, DC 20005). *Nature* specialises in the early publication of reports of scientific advances; each issue contains approximately ten longer articles, with the remainder and major part of the journal consisting of letters from scientists giving specific information on their current research work. Early publication in *Nature* has long been recognised as a method of establishing priority in a particular field. *Science* has been published since 1880 and, like *Nature*, claims to be a forum for the presentation and discussion of important items relating to the advancement of science, often including minority or conflicting views rather than the consensus. Each issue contains four or five longer articles with the bulk of the text again consisting of shorter communications. The *Franklin Institute journal* (Franklin Institute, Philadelphia, Pennsylvania 19103) is a monthly journal which was first published in 1826, and which includes original research papers in all traditional branches of mathematics and the pure and applied physical sciences, in addition to the interdisciplinary fields or the composite sciences which combine two or more disciplines.

An important popular periodical is the weekly *New scientist* (Cromwell House, Fulwood Place, London WC1.) which has appeared since 1956. The content is geared to appeal to the interested layman, in addition to the professional scientist and engineer, and the industrial applications of science are strongly featured. The remaining periodicals of importance covering science in general are:

Advancement of science. 1939-. M. British Association for the Advancement of Science, 3 Sanctuary Buildings, 20 Great Smith

Street, London SW1. A journal whose basic purpose is to inform scientists and engineers of advances of importance in disciplines other than their own.

Endeavour. 1942-. Q. Imperial Chemical Industries Ltd, Millbank, London SW1. A quarterly review of progress in science with particular emphasis on British science.

Science and technology for technical men in management. 1962-. M. Conover-Mast Publications Inc, 205 East 42nd Street, New York, NY 10017. State of the art articles on interdisciplinary subjects.

Science journal. 1965-. M. IPC Business Press Ltd, Dorset House, Stamford Street, London EC2. News and comment on science and technology in addition to state of the art surveys of specific topics.

Science progress. 1906-. Q. Blackwell Scientific Publications, Oxford. A quarterly review journal containing reviews of progress in the pure sciences.

Scientific American. 1845-. M. Scientific American, 415, Madison Avenue, New York, NY 10017.

PHYSICS

Papers on the physical basis of mechanical engineering are scattered through a large number of primary journals. The important titles covering applied physics are:

Academy of sciences of the USSR. Proceedings. Applied physics section. English translation of *Akademiia Nauk SSSR. Doklady.* 1957-. M. Plenum Publishing Corporation, 227 West 17th Street, New York, NY 10011. Cover-to-cover translation.

British journal of applied physics. Journal of physics, D. 1950-. M. Institute of Physics and the Physical Society, 47 Belgrave Square, London SW1. Original papers on the new applications of basic physical principles and research notes.

Contemporary physics. 1959-. Bi-M. Taylor and Francis Ltd, Red Lion Court, Fleet Street, London EC4. Review articles covering physics and associated technologies.

Journal of engineering physics. English translation of *Inzhenerno fizicheskii zhurnal* 1966-. M. Faraday Press, 15 Park Row, New York, NY 10038.

Journal of applied physics. 1931-. M. American Institute of Physics, 335 East 45th Street, New York, NY 10017. Original papers covering general physics and its applications to engineering and industry.

Philosophical magazine. 1791-. M. Taylor and Francis Ltd, Red Lion Court, Fleet Street, London EC4. Original papers in theoretical and applied physics.

Royal Society of London. Proceedings A : mathematical and physical sciences. 1905-. The Society, 6 Carlton House Terrace, London SW1.

MECHANICS

The significant primary journals covering the various branches of mechanics and mechanical engineering science are :

Acta mechanica. 1965-. 3 per year. Springer-Verlag, Molkerbastei 5, PO Box 367, A-1011, Wein, Austria. Text in English and German.

Archiwum mechaniki stosowanej: archives de mécanique appliquée. 1948-. Bi-M. Institut Podstawowych Problemow Techniki, Polskiej Akademii Nauk, Warsaw, Poland.

Ingenieur-archiv: gesellschaft für angewandte mathematik und mechanik. 1929-. 1 volume per year. Springer-Verlag, Heidelberger Platz 3, Berlin 31, Germany. Text in English and German.

International journal of engineering science. 1963-. Bi-M. Pergamon Press, Headington Hill Hall, Oxford. Text in English, French and German.

International journal of mechanical sciences. 1959-. M. Pergamon Press, Headington Hill Hall, Oxford.

International journal of non-linear mechanics. 1966-. M. Pergamon Press, Headington Hill Hall, Oxford.

International journal of solids and structures. 1965-. M. Pergamon Press, Headington Hill Hall, Oxford.

Journal de mécanique. 1962-. Q. Gauthiers-Villars, 55 Quai des Grands-Augustin, Paris 6e. Text in English, French or German.

Journal of applied mechanics and technical physics. 1965-. Bi-M. English translation of *Zhurnal prikladnoi mekhanik i tekhnicheskoi fiziki.* Faraday Press, 15 Park Row, New York, NY 10038.

Journal of fluid mechanics. 1956-. M. Cambridge University Press, 200 Euston Road, London NW1.

Journal of mechanical engineering science. 1959-. 5 per year. Institution of Mechanical Engineers, 1 Birdcage Walk, London SW1.
Journal of the mechanics and physics of solids. Bi-M. Pergamon Press, Headington Hill Hall, Oxford.
Physics of fluids. 1958-. M. American Institute of Physics, 335 East 45th Street, New York, NY 10017.
Quarterly journal of mechanics and applied mathematics. Q. Oxford University Press, 37 Dover Street, London W1.
Revue Roumaine des sciences techniques. Serie de mécanique appliquée. 1956-. Bi-M. Academiei RPR, Str Gutenberg 3 bis, Bucharest, Roumania. Text in English, French, German and Russian.
Soviet fluid mechanics. English translation of *Izvestiya akademiia nauk SSSR. Mekhanika zhidkostei i gasov.* 1965-. Bi-M. Faraday Press, 15 Park Road, New York, NY 10038.

ENGINEERING

There are many periodicals covering all fields of engineering: the journals issued by the general engineering societies and academies include original papers covering all facets of engineering, while journals published by commercial publishers include practical feature articles and news items applicable to many industries.

Two of the most useful news journals for the British practising engineer are *Engineer* and *Engineering*. Both these weekly periodicals (whose history has been previously noted) cover the engineering industries in general and include information on new techniques, principles and plant, in addition to features on related subjects such as industrial relations. The *Engineer* (Morgan Brothers (Publishers) Ltd, 28 Essex Street, London WC2) is mainly directed towards engineering management, with particular emphasis on design, production, factory management and research and development. *Engineering* (Engineering Ltd, 36 Bedford Street, London WC2) includes in each issue an *Engineering outline,* presenting in digest form information on such subjects as automatic process control and quality and reliability, and including references to centres of research or further information. Another useful feature of *Engineering* is the weekly ' On the shelf ' column, which draws attention to the type of engineering publication which is so often overlooked by national and trade bibliographies.

Engineer's digest (120 Wigmore Street, London W1) is a monthly, published since 1940, which is heavily slanted towards mechanical engineering and is arranged under headings such as hydraulics and pneumatics, metal finishing, etc. Each selection includes one main article with strong practical implications, while the remaining material within each section covers new developments and components. The *Engineer's digest surveys* are practical state-of-the-art accounts of subjects such as numerically controlled machine tools or drop forging. *Engineering news* (Engineering, Chemical and Marine Press Ltd, 33-39 Bowling Green Lane, London EC1) is a weekly in newspaper format which has been published since 1961 and which gives topical information on current practice in engineering design and production. *Machinery Lloyd* (Tothill Press Ltd, 161-166 Fleet Street, London EC4), which dates from 1929, is a fortnightly review of new equipment for the mechanical, electrical and construction industries.

The monthly American periodical *Power* (McGraw-Hill Inc, 330 West 42nd Street, New York, NY 10036) covers the design, operation and maintenance of power systems; in addition to practical feature articles and news of new equipment and plant, each issue includes a data sheet, past examples of which have covered the density of compressed dry air, and gas compression temperatures.

VDI zeitschrift is issued three times per month by the German Association of Engineers (Verein Deutscher Ingenieur, VDI-Verlag GmbH, Postfach 10250, 4 Düsseldorf 10, Germany). In addition to articles covering trends and developments in the various fields of engineering, and in particular production engineering, each issue includes an annual state-of-the-art survey of a selected topic, detailed abstracts of selected German engineering theses, and a classified listing of engineering articles included in over 200 German and foreign engineering journals.

Other important and mainly primary journals covering engineering in general are:

Academie Polonaise des Sciences. Bulletin. Serié des sciences techniques. 1953-. M. Export and Import Enterprise ' Ruch ', ul Wilcza 46, Warsaw, Poland. Text in English, French, German, Russian and Polish.

Acta technica (Academiae Scientarium Hungaricae). 1941-. Q. Hungarian Academy of Sciences, Alkotmany u 21, Budapest V, Hungary. Text in English, French, German and Russian.

Acta technica CSAV. 1956-. Bi-M. Ceskoslovenska Akademie Ved v Praze, Ve Smeckách 30, Prague 1, Czechoslovakia. Text in English, German and Russian.

British engineer. 1962-. Bi-M. Institution of British Engineers, Windsor House, 46 Victoria Street, London SW1.

Czechoslovak heavy industry. 1955-. M. RAPID, 13 ul 28 Rijna, Prague 1, Czechoslovakia.

Engineering journal. 1918-. M. Engineering Institute of Canada, 2050 Mansfield Street, Montreal 2, Canada.

Forschung im ingenieurwesen: zeitschrift fur technischewissenschaftliche forschung in theorie und praxis. 1930-. Bi-M. Verein Deutscher Ingenieur, VDI-Verlag GmbH, Postfach 1139, 4 Düsseldorf 1, Germany.

Industrial research. 1957-. M. Industrial Research Inc, Industrial Research Building, Beverly Shores, Indiana 46301. Controlled circulation.

Ingénieur. 1915-. M. Association des Diplomes de Polytechniques, 2500 Marie Guyard Avenue, Montreal 26, Canada. Text in French.

Institution of Engineers, Australia. Journal. 1929-. 8 per year. The Institution, Science House, Gloucester and Essex Streets, Sydney, Australia.

Journal of research. Section C: engineering and instrumentation. 1959-. Q. National Bureau of Standards. Superintendent of Documents, Washington, DC 20402.

Junior Institution of Engineers. Journal. 1891-. M. The Institution, 30 Ovington Square, London SW3.

Mining, electrical and mechanical engineer. 1920-. M. Association of Mining, Electrical and Mechanical Engineers, 62 Talbot Road, Manchester 16.

New Zealand engineering. 1946-. M. New Zealand Institution of Engineers. Technical Publications Ltd, CPO 3047, Wellington, New Zealand.

Philips technical review. 1936-. M. Centrex Publishing Co, Eindoven, Netherlands.

Russian engineering journal. English translation of *Vestnik mashinostroeniia.* 1959-. M. Production Engineering Research Association, Melton Mowbray, Leicestershire. Cover-to-cover translation.

Schweizer Archiv für Angewandte Wissenschaft und Technik: annales suisses des sciences appliquées et de la technique. 1935-. M. Bd Buchdruckerei und Verlag, Vogt-Schild AG, Solothurn, Switzerland. Text in English, French and German.

Société Royale Belge des Ingénieurs et des Industriels. Revue. 1955-. 10 per year. Hotel Ravenstein, 3 Rue Ravenstein, Brussels 1, Belgium. Text in French and other languages.

Society of Engineers. Journal and proceedings. 1854-. Q. The Society, Abbey House, Victoria Street, London SW1.

Soviet engineering journal. English translation of *Inzhenernii zhurnal.* 1965-. Bi-M. Faraday Press, 15 Park Row, New York, NY 10038. Cover-to-cover translation.

Swiss technics. 1944-. 3 per year. Swiss Office for Development of Trade, Swiss Association of Machinery Manufacturers, Av Bellefontaine 18, Lausanne, Switzerland. Editions in English, French, German, Portuguese and Spanish.

Technik: technisch-wissenschaftliche zeitschrift für grundsatz-und querschnittsfragen. 1946-. M. VEB Verlag Technik, Oranienburger Str 14-14, 102 Berlin, Germany.

Technische mitteilungen. 1920-. M. Vulkan-Verlag, Haus der Technik, 43 Essen, Germany.

Tohoku University. Technology reports. 1920-. Irreg. Faculty of Engineering, Tohoku University, Sendai, Japan.

House journals

Aiton review. 1947-. Q. Aiton & Co Ltd, Stores Road, Derby, England.

Allis-Chalmers engineering review. 1936-. Q. Allis-Chalmers Manufacturing Co, Box 512, Milwaukee, Wisconsin 53201.

Anvil. 1947-. Q. Davy-Ashmore Ltd, Darnall Works, Sheffield 9.

Brown-Boveri review. 1914-. M. British Brown-Boveri Ltd, 75 Victoria Street, London SW1.

Forge. 1958-. Q. J Brockhouse & Co Ltd, 25 Hanover Square, London W1.

Hoyt notched ingot: an engineering review and miscellany. 3 per year. Hoyt Metal Co Ltd, Berdar Road, London SW15.

Impulse. 1957-. Q. Mitchell Engineering Co Ltd, 1 Bedford Square, London WC1.

Journal of science and technology. 1934-. Q. General Electric and English Electric Co Ltd, PO Box 120, 1 Stanhope Gate, London W1.

Mobil industrial review. Irreg. Mobil Oil Co Ltd, Caxton House, Westminster, London SW1.

Parsons journal. 2 per year. C A Parsons & Co Ltd, Heaton Works, Shields Road, Newcastle-upon-Tyne 6.

Simon engineering review. 1962-. 2 per year. Simon Engineering Ltd, Cheadle Heath, Stockport, Cheshire, England.

Vickers magazine. Q. Vickers Ltd, PO Box 177, Vickers House, Melbank Tower, London SW1.

Vigilance: the quarterly journal of the National and Vulcan Engineering Insurance Group. Q. National and Vulcan Engineering Insurance Group, 14 St Mary's Parsonage, Manchester 3.

Villiers magazine. Q. Villiers Engineering Co Ltd, Marston Road, Wolverhampton, Staffordshire.

MECHANICAL ENGINEERING

The *Chartered mechanical engineer* and *Mechanical engineering* are the respective British and American periodicals which mechanical engineers use to keep abreast of professional activity. These journals, particularly the former, do not normally publish original papers; these appear in the *Proceedings of the Institution of Mechanical Engineers* and the *Transactions of the American Society of Mechanical Engineers.* The *Proceedings* have been published since 1847 and are currently issued in three parts. Part 1 consists of separately published papers, each of which is sponsored by a particular group of the Institution. Part 2A is the *Proceedings of the Institution of Mechanical Engineers, Automobile division;* these are also issued as separates and have been published in this series since 1947 when the Institution of Automobile Engineers was absorbed by the Mechanicals. Part 3 are the *Conference proceedings* which include the text of papers presented at the symposia on particular topics organised by the institution from time to time.

The *Transactions of the American Society of Mechanical Engineers,* which have been published since 1880, were reorganised in 1959 and have from this date been issued as five quarterly journals with a sixth added from 1967:

Journal of engineering for power (Series A).
Journal of engineering for industry (Series B).

Journal of heat transfer (Series C).

Journal of basic engineering (Series D).

Journal of applied mechanics (Series E). This journal has a separate history having appeared under this title since 1933 even though the papers it published subsequently appeared in the *Transactions*.

Journal of lubrication technology (Series F). 1967-.

Each of the quarterlies publish both original papers, with an average length of six thousand words, and technical briefs which usually record work in progress. The *Transactions* are available as individual quarterlies or as an annual set in three hardbound volumes—vol 1 covering series A/B; vol 2, C, D, F; vol 3, E.

Other periodicals covering mechanical engineering are:

Acta polytechnica Scandinavica. Mechanical engineering series. 1958-. Irreg. Scandinavian Council for Applied Research, Box 5073, Stockholm 5, Sweden.

Archiwum budowy maszyn: the archive of mechanical engineering. 1954-. Q. Polska Akademia Nauk, Komitet Budowy Maszyn, ul Nowowiejska 25, Warsaw, Poland. Text in English, French, German or Polish.

Australasian engineer. 1909-. M. Australian Institute of Metals, Society of Mechanical Engineers of Australia, 116-126 Cleveland Street, Chippendale, New South Wales, Australia.

Bulletin of mechanical engineering education. 1962-. Q. University of Manchester Institute of Science and Technology. Pergamon Press, Headington Hill Hall, Oxford.

Institution of Engineers, Australia. Mechanical and chemical engineering transactions. 1965-. 2 per year. The Institution, Science House, Gloucester and Essex Streets, Sydney, Australia.

Institution of Engineers (India). Journal. Mechanical engineering division. 1920-. Bi-M. The Institution, 8 Gokhale Road, Calcutta 20, India.

Japan Society of Mechanical Engineers. Bulletin. 1958-. Bi-M. Japan Publications Trading Co Ltd, CPO Box 722, Tokyo, Japan. Text in English and German.

Konstruktion im maschinen-, apparate-und gerätebau. 1949-. M. Springer-Verlag, Heidelberger Platz 3, 1 Berlin 31, Germany.

Maschinenbautechnik. 1952-. M. VEB Verlag Technik, Oranien-burger Str 1-14, 102 Berlin, Germany.

Mechanical engineering news. 1964-. Q. Mechanical Engineering Division, American Society of Engineering Education. Oklahoma State University, Department of Mechanical Engineering, Stillwater, Oklahoma.

Mechanik miesiecznik naukowo-techniczny. 1909-. M. Wydawnictwa Czasopism Techn NOT, Czackiego 3/5, Warsaw, Poland. Contents page in Czech, English, German and Polish.

National engineer. 1897-. M. National Association of Power Engineers Inc, 176 West Adams Street, Chicago, Illinois 60603.

Polytechnisch tijdschrift, edition A: Mechanical engineering, machinery, shipbuilding. 1946-. 6 per year. Technische Uitgeverij H Stam NV, Industrieweg 1, Culemborg, Netherlands.

Power transmission design: motors, drives, bearings and controls. 1959-. M. Industrial Publishing Corp, 812 Huron Road, Cleveland, Ohio 44115.

Pratique des industries mécaniques. 1913-. M. DUNOD, 92 Rue Bonaparte, Paris 6°.

Revue mécanique. 1954-. Q. Société Belge des Mécaniciens, 21 Rue des Drapiers, Brussels, Belgium.

South African mechanical engineer. 1951-. M. South African Institution of Mechanical Engineers, Kelvin Publications Ltd, London House, Loveday Street, Johannesburg, South Africa. Text in Afrikaans and English.

House journals

Apex. 1969-. Q. Amalgamated Power Engineering Ltd, Bedford, England.

Bristol Siddeley journal. 1960-. Q. Bristol Siddeley Engines Ltd, Mercury House, 195 Knightsbridge, London SW7.

Escher Wyss news. Q. Escher Wyss Ltd, 8023 Zurich, Switzerland.

Fairey review. 1958-. Q. Fairey Co Ltd, 24 Bruton Street, London W1.

Leyland journal. 1935-. Q. Leyland Group, Leyland, Lancashire, England.

Power specialist. 1924-. Q. Johns-Manville Corporation, 22 East 40th Street, New York, NY 10016.

Precision: devoted to the interests of the automotive, aircraft, general engineering and allied industries. 1946-. M. Automotive Products Group Ltd, Leamington Spa, England.

Sulzer technical review. 1920-. Q. Sulzer Brothers Ltd, Winterthur, Switzerland. Editions in English, French, German, Portuguese and Spanish.

V-belt journal. Q. J H Fenner & Co Ltd, Marfleet, Hull, Yorkshire.

The following selection of periodicals, arranged within one alphabetical sequence of subject headings, covers both journals on specific subjects within the field of mechanical engineering, and also journals covering related subjects and industries. In ' core ' subjects, for example, fabrication and joining, and tribology, an attempt has been made to include every important English language journal on the subject and the most important foreign language titles. Coverage of related subjects and industries, for example, plastics and marine engineering, is not as exhaustive as that of the ' core ' subjects in that only the outstanding titles in English and other languages are included. In all cases only those Russian journals which have been translated into English in the form of cover-to-cover translations have been included.

AERONAUTICAL AND AEROSPACE ENGINEERING

Aeronautical journal. 1897-. M. Royal Aeronautical Society, 4 Hamilton Row, London W1.

Aeronautical quarterly. 1949-. Q. Royal Aeronautical Society, 4 Hamilton Place, London W1.

AIAA journal. 1963-. M. American Institute of Aeronautics and Astronautics, 1290 Avenue of the Americas, New York, NY 10019.

American aviation. 1937-. M. American Aviation Pub Inc, 1000 Vermont Avenue, Washington, DC 20005.

Aviation and cosmonautics. English translation of *Aviatsiia i kosmonavtika. irreg.* Foreign Technology Division, Air Force Systems Command, Wright-Patterson Air Force Base, Ohio. Cover-to-cover translation.

Aviation week and space technology. 1916-. W. PO Box 430, Hightstown, New Jersey 08520.

Flight international. 1909-. W. Iliffe Transport Publications Ltd, Dorset House, Stamford Street, London SE1.

Flug revue jetzt vereinigt mit flugwelt. M. Vereinigte Motor-Verlage GmbH, 7 Stuttgart I, Motor-presse Haus, Leuschnerstrasse I, Postfach 1042.

Helicopter world. 1958-. M. Air Age Publications Ltd, 1 Temple Chambers, Temple Avenue, London EC4.

Hovercraft world. 1966-. 8 per year. Air Age Publications, 1 Temple Chambers, Temple Avenue, London EC4.

Hoveringcraft and hydrofoil: the international review of air cushion vehicles and hydrofoils. 1961-. M. Kalerghi Publications, 50/52 Blandford Street, London W1.

Journal of aircraft. 1964-. Bi-M. American Institute of Aeronautics and Astronautics, 1290 Avenue of the Americas, New York, NY 10019.

Journal of spacecraft and rockets. 1964-. Q. American Institute of Aeronautics and Astronautics, 1290 Avenue of the Americas, New York, NY 10019.

Reed's aircraft and equipment news. 1959-. M. Thomas Reed Publications Ltd, 38 St Andrews Hill, London EC4.

SAE journal. 1917-. M. Society of Automotive Engineers, 485 Lexington Avenue, New York, NY 10017.

Soviet aeronautics. English translation of *Izvestiya vysshikh uchebnykh zavedenii aviatsionnaya teknika.* 1966-. Q. Faraday Press Inc, 84 Fifth Avenue, New York, NY 10011. Cover-to-cover translation.

Tech air. 1945-. M. Society of Licensed Aircraft Engineers and Technologists, 153 London Road, Kingston-upon Thames, England.

Vertical world: magazine of helicopters. 1965-. Bi-M. Vertical World Inc, Suite 407, 1317 F Street, Washington, DC 20004.

Zeitschrift fur flugwissenschaften. 1953-. M. Friedr Vieweg & Sohn GmbH, Postfach 185, 33 Braunschweig, Germany.

House journals

Air BP: journal of the international aviation service of the BP group. Q. British Petroleum Co Ltd, Britannic House, Moor Lane, London EC2.

Esso air world. 1947-. Bi-M. Esso International Inc, 15 West 51st Street, New York, NY 10019.

Hawker Siddeley technical review. 1958-. Q. Hawker Siddeley Group, Duke's Court, 32 Duke Street, St James', London SW1.

Lockheed horizons. Irreg. Lockheed-California Co, Burbank, California 91503.

Shell aviation news. 1931-. M. Shell Aviation News, Shell Centre, London SE1.

SI aviation review. 1955-. Irreg. Smith's Industries Ltd, Kelvin House, Wembley Park Drive, Wembley, Middlesex.

AIR CONDITIONING see HEATING AND VENTILATION ENGINEERING

ALUMINIUM

Aluminium: fachzeitschrift der aluminium-industrie. 1917-. M. Aluminium-Zentrale e/v, Aluminium-Verlag GmbH, Jägerhofstr 29, Postfach 10008, 4000 Düsseldorf, Germany.

Revue de l'aluminium et de ses applications. 1924-. M. Société d'Edition et de Documentation des Alliages, 3 rue Saint-Philippe-du-Roule, Paris 8ᵉ.

Schweizer aluminium rundschau/Revue suisse de l'aluminium. 1951-. 9 per yr. Interessengemeinschaft de Schweiz, Aluminium-Hütten, Walz- und Presswerke, Utoquai 37, CH-8008, Zurich, Switzerland.

House journals

Alcan magazine. 1951-. Q. Alcan Industries Ltd, Banbury, Oxfordshire, England.

Kaiser aluminum news. 1942-. Bi-M. Kaiser Aluminum & Chemical Corp, Kaiser Center, 300 Lakeside Drive, Oakland, California 90012.

APPLIED HEAT

Brennstoff—wärme—kraft BWK: zeitschrift für energietechnik und energiewirtschaft. 1949-. M. Verein Deutscher Ingenieure, VDI-Verlag GmbH, 4 Düsseldorf 1, Postfach 1139, Germany.

Combustion and flame. 1957-. Q. Combustion Institute, Butterworth & Co (Pubs) Ltd, 88 Kingsway, London WC2.

Combustion, explosion and shock waves. English translation of *Fizika goreniia i vzryva.* 1965-. Q. Faraday Press, 15 Park Row, New York, NY 10038. Cover-to-cover translation.

Entropie: revue scientifique et technique de thermodynamique. 6 per year. Editions Bartheye, 54 Avenue Marceau, Paris 8ᵉ.

Fuel: a journal of fuel science. 1922-. Bi-M. Butterworth & Co (Pubs) Ltd, 88 Kingsway, London WC2.

Institute of Fuel journal. 1926-. M. Institute of Fuel, 18 Devonshire Street, Portland Place, London WI.

International journal of heat and mass transfer. 1960-. M. Pergamon Press, Headington Hill Hall, Oxford.

Journal of heat transfer (*Transactions of the American Society of Mechanical Engineers, series C*) see page 116.

Revue générale de thermique: combustibles énergie, équipéments thermiques. 1962-. M. Georges Nerot, 2 rue des Tanneries, Paris 13ᵉ.

Thermal engineering. English translation of *Teploenergetika*. M. Pergamon Press, Headington Hill Hall, Oxford. Cover-to-cover translation.

Wärme-und stoffübertragung. 1968-. Q. Springer Verlag. I Berlin 33, Heidelberger Platz 3, Germany.

See also ENGINE DESIGN AND APPLICATION; HEATING AND VENTILATING ENGINEERING; REFRIGERATION; ENGINEERING; STEAM ENGINEERING

ASSEMBLY ENGINEERING see FABRICATION AND JOINTING

ASTRONAUTICS see AERONAUTICAL AND AEROSPACE ENGINEERING

AUTOMATION AND INSTRUMENTATION

Automatic control. English translation of *Avtomatika i vychislitel' naya technika*. 1967-. M. Faraday Press Inc, 84 Fifth Avenue, New York, NY 10011.

Automatica: the international journal on automatic control and automation. 1963-. Q. Pergamon Press, Headington Hill Hall, Oxford. Text in English, French, German and Russian.

Automation: the magazine of profitable production. 1954-. M. Penton Pub Co, Penton Building, Cleveland, Ohio. Controlled circulation.

Automation and remote control. English translation of *Avtomatika i telemekhanika*. 1956-. M. Plenum Publishing Corporation, 227 West 17th Street, New York, NY 10011. Cover-to-cover translation.

Automatisierung. 1956-. M. Karl M Hagenerer-Verlag, Blumenthalstr, 40, Postfach 868, 69 Heidelberg 1, Germany.

Automatisme. 1956-. M. Association Française de Régulation et d'Automatisme, 92 rue Bonaparte, Paris⁶.

Canadian controls and instrumentation. 1962-. M. Maclean-Hunter Pub Co Ltd, 481 University Avenue, Toronto 2, Canada.

Control and automation progress. 1958-. M. Morgan Bros (Publishers) Ltd, 28 Essex Street, Strand, London WC2.

Control engineering. 1954-. M. McGraw-Hill Inc, 466 Lexington Avenue, New York, NY 10017. Controlled circulation.

Information and control. 1957-. Bi-M. Academic Press Inc, 111 Fifth Avenue, New York, NY, 10003.

Instrument and control engineering. 1963-. M. Tothill Press Ltd, 161/166 Fleet Street, London EC4. Controlled circulation.

Instrument practice. 1946-. M. United Trade Press Ltd, 9 Gough Square, London EC4.

Instrument technology. 1954-. M. Instrument Society of America, 530 William Penn Place, Pittsburgh, Pennsylvania 15219.

Instruments and control systems. 1928-. M. Rimbach Publications, 845 Ridge Avenue, Pittsburgh, Pennsylvania 15212.

Journal of scientific instruments. Journal of physics E. 1922-. M. Institute of Physics and the Physical Society, 1 Lowther Gardens, Prince Consort Road, London SW7.

Laboratory equipment digest. 1963-. M. Gerard Mann Ltd, 1-3 Astoria Parade, Streatham High Road, London SW16.

Laboratory practice: the journal for techniques and equipment in all branches of science. 1952-. M. United Trade Press, 9 Gough Square, Fleet Street, London EC4.

Review of scientific instruments. 1930-. M. American Institute of Physics, 335E 45th Street New York NY 10017.

Society of Instrument Technology. Transactions. 1949-. Q. Society of Instrument Technology, 20 Peel Street, London W8.

Soviet journal of instrumentation and control. English translation of *Pribory i sistemy upravleniia.* 1967-. Scripta Technica, 138 New Bond Street, London W11. Cover-to-cover translation.

Steuerungs-technik international: machine control, production control, application of NCMT. 1968-. M. Kransskopf-verlag für Wirtschaft GmbH & Co, 65 Mainz, Lessingstrasse 12-14, Germany.

Zeitschrift für instrumentenkunde. 1881-. M. Verlag Friedr Vieweg & Sohn GmbH, Burgplatz 1, Braunschweig, Germany.

House journals

Hilger journal. 1954-. Q. Hilger & Watts Ltd, 98 St Pancras Way, London, NW1.

Instrument engineer. 1952-. 2 per year. George Kent Ltd, Biscot Road, Luton, Bedfordshire.

Instrumentation. 1947-. Q. Honeywell Controls Ltd, Ruislip Road East, Greenford, Middlesex.

Muirhead technique: a journal of instrument engineering. 1947-. Q. Muirhead & Co Ltd, Beckenham, Kent.

Pulse. 1959-. Irreg. Smith's Industrial Division, Kelvin House, Wembley Park Road, Wembley, Middlesex.

See also METROLOGY.

AUTOMOBILE ENGINEERING

ATZ-automobiletechnische zeitschrift. 1898-. M. Franckh'sche Verlagshandlung, Pfizerstr 5, Stuttgart 0, Germany.

Autocar. 1895-. W. Iliffe Transport Publications Ltd, Dorset House, Stamford Street, London SE1.

Automobile engineer: design, material production methods and works equipment. 1910-. M. Iliffe Transport Publications Ltd, Dorset House, Stamford Street, London SE1.

Automotive design engineering: a technical monthly for designers of cars, vans, trucks, tractors, buses, etc. 1962-. M. Hermes House, Blackfriars Road, London SE1. Controlled circulation.

Automotive industries. 1899-. Semi-M. Chilton Co, Chestnut and 56th Streets, Philadelphia, Pennsylvania 19139.

IAAE journal. 1942-. M. Institution of Automotive and Aeronautical Engineers, Kelvin Hall, 55 Collins Place, Melbourne C1, Australia.

Ingénieurs de l'automobile. 1927-. M. Société des Ingénieurs de l'Automobile, 254 Rue de Vangirad, 75 Paris 15e.

Institute of Road Transport Engineers. Journals and proceedings. 1945-. Q. The Institute of Road Transport Engineers, 1 Cromwell Place, London SW7.

Institution of Mechanical Engineers. Proceedings. Automobile division. 1947-. Irreg. The Institution of Mechanical Engineers, 1 Birdcage Walk, London SW1.

Motor. 1903-. W. Temple Press Ltd, Bowling Green Lane, London EC1.

MTZ-Motortechnische zeitschrift. 1938-. M. Franckh'sche Verlagshandlung W Keller & Co, Pfizstr 5-7, Stuttgart 1, Germany.

SAE journal. 1917-. M. Society of Automotive Engineers, 485 Lexington Avenue, New York, NY 10017. In addition to the papers included in the journal each year, some 500 technical papers are issued on current development in ground vehicle engineering and aerospace. These may be ordered individually at $1.00 each, or standing orders may be placed for a complete set of the papers, or for those covering ground vehicle engineering or aerospace. *SAE transactions* is an annual publication of 5 volumes including 200 selected papers. A complete checklist of SAE papers is issued 3 times per year.
House journal

AC-Delco news. 2 per year. AC-Delco Division, General Motors Ltd, Dunstable, Bedfordshire.

BEARINGS
House journals
Ball bearing journal: a quarterly review of rolling bearing engineering. 1925-. Q. Skefko Ball Bearing Co Ltd, Luton, Bedfordshire.

Bearing engineer. 1940-. Bi-M. Torrington Co, Bearings Division, Torrington, Connecticut.

Kugellager-zeitschrift: fackzeitschrift für kugellager und rollenlager. Q. Aktiebolaget Svenska, Kuelagerfabriken, Goteborg, Sweden.

Motion research and engineering. SKF Industries Inc, Front Street and Erie Avenue, Philadelphia 32, Pennsylvania.

Timken. M. Timken Roller Bearing Co, Dunston, Northampton.

See also TRIBOLOGY

CASTING see FOUNDRY PRACTICE

COBALT
House journal
Cobalt. 1958-. Q. Centre d'Information du Cobalt, 35 Rue des
Colonies, Brussels, Belgium. Editions in English, French and German.

COMBUSTION see APPLIED HEAT

COMPRESSED AIR see FLUID POWER AND PNEUMATICS

CONTROL ENGINEERING see AUTOMATION AND INSTRUMENTATION

COPPER
Cuivre, laitons, alliages: revue technique et de vulgarisation. 1951-.
5 per year. Editions Techniques Riegel, 45 Avenue du Roule, Neuilly-
sur-Seine, France.
House journals
Canadian coppermetals. 1960-. Q. Canadian Copper and Brass
Development Association, 55 York Street, Toronto 1, Canada.
Copper. 1958-. 3 per year. Copper Development Association, 55
South Audley Street, London W1.

CORROSION
Anti-corrosion methods and materials: the first British journal of
corrosion control, prevention, engineering and research. 1954-. M.
Grampian Press Ltd, The Tower, 229-243 Shepherds Bush Road,
Hammersmith, London W6.
Australasian corrosion engineering: a technical journal for the
control and prevention of corrosion. 1957-. M. Box 250, North Sydney,
NSW, Australia.
British corrosion journal. 1965-. Bi-M. British Joint Corrosion
Group, 14 Belgrave Square, London SW1.
Corrosion: a journal of science and engineering. National Associa-
tion of Corrosion Engineers, 980 M & M Building, Houston, Texas
77002.
Corrosion et anti-corrosion. 1953-. M. Société de Productions Docu-
mentaires, 28 Rue St-Dominique, Paris 7e.
Corrosion prevention and control. 1954-. M. Scientific Surveys Ltd.
11A Gloucester Road, London SW7.

Corrosion science. 1961-. M. Pergamon Press, Headington Hill Hall, Oxford.

Materials protection. 1962-. M. National Association of Corrosion Engineers, 980 M & M Building, Houston, Texas 77002.

Protection of metals. English translation of *Zashchita metallov* 1962-. M. Scientific Information Consultants Ltd, 661 Finchley Road, London NW2. Cover-to-cover translation.

Werkstoffe und korrosion. 1950-. M. Verlag Chemie CmbH, Weinheim, Bergstr, Germany.

House journal

PIQ (process industries quarterly). 1939-. Q. George L Lansdale, Huntington Alloy Products Division, International Nickel Co Inc, Huntington, W Virginia 25720.

CRYOGENICS see REFRIGERATION ENGINEERING
DIESEL ENGINES see ENGINE DESIGN AND APPLICATION
ELECTROPLATING see METAL FINISHING

ENGINE DESIGN AND APPLICATION

Diesel and gas turbine progress. 1935-. M. 11225 W Bluemound Road, Box 7406, Milwaukee, Wisconsin 53213.

Diesel equipment superintendent: the maintenance management magazine serving the man responsible for selection, performance and availability of diesel equipment. 1923-. M. Diesel Publications Inc, 80 Lincoln Avenue, Stamford, Connecticut 06902.

Engine design & applications. 1964-. M. Harlequin Press Ltd, Grand Buildings, Trafalgar Square, London WC2.

Gas & oil power. 1905-. M. Whitehall Technical Press Ltd, Wrotham Place, Sevenoaks, Kent.

Gas turbine international. 1959-. Bi-M. Gas Turbine Publications Inc, 80 Lincoln Avenue, Stamford, Connecticut 06902.

Mitteilungen aus dem institut für thermische turbomaschinen. 1950-. Irreg. Juris Verlag, Zurich, Switzerland.

House journals

Drive. Q. Ruston & Hornsby Ltd, Sheaf Iron Works, Lincoln.

Heat engineering. 1926-. Foster Wheeler Corp, Livingston, New Jersey 07039.

Oil engine news. 1959-. Irreg. Marketing Dept, Oil Engine Division, Rolls Royce Ltd, Shrewsbury.

Rolls-Royce journal. 1968-. Q. Rolls Royce Ltd., Derby.

See also AERONAUTICAL AND AEROSPACE ENGINEERING; APPLIED HEAT; AUTOMOBILE ENGINEERING; LOCOMOTIVE ENGINEERING; MACHINE DESIGN; MARINE ENGINEERING

ENGINEERING DESIGN

Design & components in engineering. 1961-. Semi-M. Tothill Press Ltd, 161/166 Fleet Street, London EC4. Controlled circulation.

Design engineering. 1955-. M. Maclean-Hunter Publishing Co, 481 University Avenue, Toronto 2, Canada.

EM & D: journal of engineering materials, components and design. 1958-. M. Heywood-Temple Industrial Publications Ltd, 33-39 Bowling Green Lane, London EC1.

Engineering designer. 1950-. M. Institution of Engineering Designers, 1 Earl's Lane, South Mimms, Potters Bar, Hertfordshire.

Ergonomics: human factors in work, machine control and equipment design. 1957-. Q. Ergonomics Research Society. Taylor & Francis Ltd, Red Lion Court, Fleet Street, London EC4.

Materials in design engineering. 1929-. M. Reinhold Publishing Co, 430 Park Avenue, New York, NY 10022.

Product design and development. 1946-. M. Chilton Co, Chestnut & 56th Streets Philadelphia, Pennsylvania 19139. Controlled circulation.

Product design and value engineering. 1956-. M. Southam Business Publications Ltd, 1450 Dons Mill Road, Ontario, Canada.

Product design engineering: a technical monthly for designers using mechanical, electro-mechanical and applied electronic techniques. 1963-. M. Design Engineering Publications Ltd, Hermes House, 89 Blackfriars Road, London SE1. Controlled circulation.

See also ENGINE DESIGN AND APPLICATION; MACHINE DESIGN

ENGINEERING INSPECTION see QUALITY CONTROL AND MEASUREMENT

ENVIRONMENTAL ENGINEERING

Environmental engineering, 1962-. Q. Society for Environmental Engineers, Kenneth Mason Publications Ltd, 14 Homewell, Havant, Hampshire.

Institute of environmental sciences. Proceedings. 1955-. Annual. Institute of Environmental Sciences 940 East Northwest Highway, Mount Prospect, Illinois 60056.

Journal of the environmental sciences. 1958-. Bi-M. Institute of Environmental Sciences, 940 East Northwest Highway, Mount Prospect, Illinois 60056.

See also MATERIALS SCIENCE AND ENGINEERING

ERGONOMICS see ENGINEERING DESIGN

FABRICATION AND JOINING

Assembly and fastener methods. 1963-. M. Grampian Press Ltd, 229-243 Shepherds Bush Road, London w6.

Assembly engineering. 1958-. M. Hitchcock Publishing Co, Geneva Road, Wheaton, Illinois 60188.

Australian welding journal. 1956-. Bi-M. Australian Welding Institute, Australian Trade Publications, 28 Chippen Street, Chippendale, NSW, Australia.

Automatic welding. English translation of *Avtomaticheskaia svarka.* 1959-. M. Welding Institute, Abington Hall, Abington, Cambridge, England. Cover-to-cover translation.

Canadian welder & fabricator. 1909-. M. Sanford Evans Publishing Ltd, Box 6900, Winnipeg 21, Manitoba, Canada.

Industrial welding and modern jointing. 1961-. M. Scientific Surveys Ltd, 11A Gloucester Road, 32 Southwark Bridge, London SE1.

Metal construction and British welding journal. 1954-. M. Welding Institute, Abington Hall, Cambridge, England.

Der praktiker schweissen und schneiden. 1949-. M. Deutscher Verlag für Schweisstechnik (DVS) GmbH, Düsseldorf, Schadowstr 42, Germany.

Revue de la soudure-lastijdschrift. 1945-. Q. Institut Belge de la Soudure, Rue des Drapiers 21, Brussels 5, Belgium. Text in Dutch and French.

Schweiss technik soudure: zeitschrift für schweisstechnik. 1910-. M. 4000 Bâle 6, St-Alban-Vorstadt, Switzerland. Text in German and French.

Schweissen und schneiden: technisch-wissenschaftliche zeitschrift des Deutschen Verbandes für Schweisstechnik. 1948-. M. Deutscher Verlag für Schweisstechnik GmbH, 4 Düsseldorf, Schadowstr 42, Germany.

Schweisstechnik: zeitschrift für alle gebiete der schweiss-, schneid- und löttetechnik. 1951-. M. VEB Verlag Technik, Oranienburger Str 13-14, 102 Berlin, Germany.

Soudage et techniques connexes. 1947-. M. Publications de la Soudure Autogene, 32 Bd de la Chapelle, Paris 18e.

Welder's world. 1964-. M. Welder's World Inc, 6006 Schroeder Road, Houston, Texas 77021.

Welding and metal fabrication. 1933-. M. Iliffe Production Publications Ltd, Dorset House, Stamford Street, London SE1.

Welding design and fabrication. 1930-. M. Industrial Publishing Co, Div of Pittsburgh Railways Co, 812 Huron Road, Cleveland, Ohio 44115.

Welding engineer. 1916-. Welding Engineer Publications Inc, Box 128, Morton Grove, Illinois 60053.

Welding in the world/Soudage dans le monde. 1963-. Q. International Institute of Welding, 54 Princes Gate, Exhibition Road, London SW7. Text in English and French.

Welding journal. 1922-. M. American Welding Society, 345 East 47th Street, New York, NY 10017.

Welding production. English translation of *Svarochnoe proizvodstvo.* 1959-. M. Welding Institute, Abington Hall, Abington, Cambridge, England. Cover-to-cover translation.

Welding Research Council bulletin. 1949-. 9 per year. Welding Research Council, 345 East 47th Street, New York, NY 10017.

House journals

Fastners. 1944-. Q. Industrial Fasteners Institute, 1505 East Ohio Building 1717 East 9th Street, Cleveland, Ohio 44114.

Fusion facts. 1927-. Q. Stoody Co, 12023 East Slauson Avenue, Whittier, California 90606.

Hobart weldword. 1940-. Q. Hobart Brothers Co, Hobart Square, Troy, Ohio 45373.

SIF-tips. 1957-. Q. Suffolk Iron Foundry (1920) Ltd, Sifbronze Works, Stowmarket, Suffolk.

Stabilizer. 1936-. Q. Lincoln Electric Co Ltd, Black Fan Road, Welwyn Garden City, Hertfordshire.

Svetsaren: ESAB in English. 1936-. Q. ESAB, Göteborg 8, Sweden.

Torch. 1954-. Q. British Oxygen Co Ltd, Bridgewater House, Cleveland Row, St James, London SW1.

Welder. 1929-. Q. Murex Welding Processes Ltd, Hertford Road, Waltham Cross, Hertfordshire.

see also WORKSHOP TECHNOLOGY

FASTENERS see FABRICATION AND JOINING

FLUID POWER AND PNEUMATICS

Compressed air: a review of the capabilities and economics of air and gases. 1896-. M. Compressed Air Magazine Co, 942 Memorial Parkway, Phillipsburg, New Jersey 08865. Editions in English, French, German, Italian and Spanish.

Fluid power international: the international journal of the hydraulic pneumatic and control industries . . . 1935-. M. Grampian Press Ltd, 229-243 Shepherds Bush Road, Hammersmith, London W6.

Fluidics international. 1968-. Q. Producer Journals Ltd, Summit House, Glebe Way, West Wickham, Kent.

Hydraulic pneumatic power. 1955-. M. Trade and Technical Press Ltd, Crown House, Morden, Surrey.

Hydraulics and pneumatics: the magazine of fluid power systems. 1948-. M. Industrial Publishing Corporation, 614 Superior Avenue West, Cleveland, Ohio 44113.

Hydraulique, pneumatique & asservissements. 1964-. 9 per year. Syndicat de Constructeurs de Transmissions Hydraulique et Pneumatiques, Compagnie Française d'Editions SA, 40 Rue du Colisée, Paris 8ᵉ.

Journal of vacuum science and technology. 1964-. Bi-M. American Vacuum Society, American Institute of Physics, 335 East 45th Street, New York, NY, 10017.

Olhydraulik und pneumatik international: unabhängige zeitschrift für kraftübertragung, regelung und steuerung. 1957-. M. KG Krausskopf-Verlag für Wirtschaft GmbH, Lessingstr 12-14, 65 Mainz, Germany.

Power futures. 1957-. Q. D M West, Box 681, Santa Ana, California 92702. Text mainly in English, occasionally in German.

Pumpen und verdichter informationen. 2 per year. VEB Pumpen-und Geblasewerk, DDR 7031 Leipzig, Klingenstr 16/18, Germany.

Pumps/Pompes/Pumpen: the international journal of pump application. 1965-. 9 per year. Trade and Technical Press Ltd, Crown House, Morden, Surrey.

Vaccum. M. Pergamon Press Ltd, Headington Hall, Oxford.

Vakuumtechnik: internationale zeitschrift für alle fragen der vakuum-technik. 1952-. M. Rudolf A. Lang. Bismarckplatz 6, 6200 Wiesbaden, Germany.

Vide: technique, applications. 1946-. BiM. Sociétié Française des Ingénieurs et Techniciens du Vide, 147 ter A, Boulevard de Strasbourg, Nogent-sur-Marne (Seine), France.

Vizugyi kozlemenyek/Hydraulic engineering. 1879-. Q. Vizgazdálkodási Tudományos Kutató Intézet, Rákoczi ut 41, Budapest VIII, Hungary.

House journals

Chesterfield news. Q. Chesterfield Tube Co Ltd, Chesterfield, Derbyshire.

Compressed air comments. 1958-. 3 per year. Atlas Copco (Great Britain) Ltd, Maylands Avenue, Hemel Hempstead, Hertfordshire.

Journal of applied pneumatics. 1952-. Q. Martonair Ltd, Twickenham, Middlesex, England.

Successful hydraulics. Q. Keelavite Hydraulics Ltd, Allesley, Coventry, Warwickshire, England.

FORGING see WORKSHOP TECHNOLOGY

FOUNDRY PRACTICE

AFS cast iron metals research journal. Q. American Foundrymen's Society, Golf and Wolf Roads, Des Plaines, Illinois 60016.

Blast furnace and steel plant. 1913-. M. Steel Publications, Grant Building, Pittsburgh, Pennsylvania 15230.

British foundryman. 1908-. M. Institute of British Foundrymen, 137-139 Euston Road, London NW1.

Castings: a journal for the foundryman. 1955-. M. F. W. Publications, 310 George Street, Sydney, Australia.

Diecasting & metal moulding. 1964-. Bi-M. Brooklands Press Ltd, Ashley House, Hatton Garden, London EC1. Controlled circulation.

Fonderie. 1946-. M. Editions Techniques des Industries de la Fonderie, 12 Avenue Raphael, Paris 16°.

Foundry. 1892-. M. Penton Publishing Co, Penton Building, Cleveland, Ohio 44113.

Foundry trade journal. 1902-. W. Fuel and Metallurgical Journals Ltd, 17 John Adam Street, Adelphi, London WC2.

Giesserei: zeitschrift für das gesamte giessereiwesen. 1914-. semi-M. Verein Deutscher Giessereifachleute und Wirtschaftsverband Giesserei-Industrie. Giesserei-Verlag GmbH, Breite Str 27, Postfach 3503, Düsseldorf, Germany.

Giessereitechnik: technische zeitschrift für das giessereiwesen. 1955-. M. VEB Deutscher Verlag für Grundstoffindustrie, Karl-Heine-Str 27, 7031 Leipzig, Germany.

Modern castings. 1938-. M. American Foundrymen's Society Inc, Golf & Wolf Roads, Des Plaines, Illinois 60016.

Precision metal: design/finishing & use of die castings; extrusion; forgings; investment castings; powder metallurgy parts; permanent & plaster & shell mold castings. 1943-. M. Industrial Publishing Corp, 812 Huron Road, Cleveland, Ohio 44114.

Russian castings production. English translation of *Liteinoe proizvodstvo.* 1961-. M. British Cast Iron Research Association, Alvechurch, Birmingham. Cover-to-cover translation.

House journals

Better methods: dedicated to continued progress in the foundry. 1926-. Q. Beardsley & Piper Division, Pettibone Mulliken Corp, 5501 West Grand Avenue, Chicago, Illinois 60639.

Butterfly foundry news. 1958-. 2 per year. Hunter and Barney Ltd, The Butterfly Co Ltd, Ripley, Derbyshire.

Die casting engineer. 1957-. Bi-M. Society of Die Casting Engineers Inc, 14530 West 8 Mile Road, Detroit, Michigan 48237.

Foseco foundry practice. 1954-. Bi-M. Foseco Inc, Box 8728, Cleveland, Ohio 44135.

See also WORKSHOP TECHNOLOGY

FRICTION see TRIBOLOGY

GAS TURBINES see ENGINE DESIGN AND APPLICATION

GRINDING see WORKSHOP TECHNOLOGY

HEAT TRANSFER see APPLIED HEAT

HEATING AND VENTILATING ENGINEERING

Air conditioning, heating and refrigeration news. 1926-. W. Business News Publishing Co, 450 West Fort Street, Detroit, Michigan 48226.

Air conditioning, heating and ventilating. 1904-. M. Industrial Press, 200 Madison Avenue, New York, NY 10016.

Allgemeine wärmetechnik: zeitschrift für wärmekälte und verfahrenstechnik. 1950-. M. Postfach 191, 623 Frankfurt-Hoechst, Germany.

ASHRAE journal: heating, refrigerating, air-conditioning, ventilating. 1904-. M. American Society of Heating, Refrigerating and Air-Conditioning Engineers, 345 East 47th Street, New York, NY 10017.

ASHRAE transactions. 1895-. 2 per year. American Society of Heating, Refrigerating and Air-Conditioning Engineers, 345 East 47th Street, New York, NY 10017.

Domestic engineering: mechanical, contracting, management, marketing. 1889-. M. Werner Ellman, Medalist Publishing Co, 1801 Prairie Avenue, Chicago, Illinois 60616. Air-conditioning.

Heating and ventilating engineer and journal of air conditioning. 1928-. M. Technitrade Journals Ltd, 11-13 Southampton Row, London WC1.

Heating and ventilating news. 1963-. M. Brooklands Press Ltd, 96 Hatton Garden, London EC1. Controlled circulation.

Heating, piping & air conditioning. 1929-. M. Keeney Division, Reinhold Publishing Corp, 10 South La Salle Street, Chicago, Illinois 60603.

Heating/plumbing/air conditioning. 1923-. semi-M. Southam Business Publications Ltd, 1450 Don Mills Road, Ontario, Canada.

Institution of heating and ventilating engineers. Journal. 1933-. M. Institution of Heating and Ventilating Engineers, Cadogan Square, London SW1.

Journal of fuel and heating technology. 1952-. Bi-M. Arrow Press Ltd, 65-66 Turnmill Street, London EC4.

Oil and gas firing. 1958-. M. Oil Firing Publications Ltd, Packhams, North Street, Midhurst, Sussex.

See also REFRIGERATION ENGINEERING

133

INDUSTRIAL AND PROCESS HEATING

Industrial and process heating. 1968-. M. Factory Publications, Hermes House, 89 Blackfriars Road, London, SE1. Controlled circulation.

Industrial heating. 1934-. M. National Industrial Publishing Co, Union Trust Building, Pittsburgh 19, Pennsylvania.

Traitement thermique/Heat treatment. 1963-. Bi-M. 254 Rue de Vaugirard, Paris 15^e.

HELICOPTERS see AERONAUTICAL AND AEROSPACE ENGINEERING

HOVERCRAFT see AERONAUTICAL AND AEROSPACE ENGINEERING

HYDRAULICS see FLUID POWER AND PNEUMATICS

INDUSTRIAL EQUIPMENT see PLANT ENGINEERING AND INDUSTRIAL EQUIPMENT

INSTRUMENTATION see AUTOMATION AND INSTRUMENTATION

INSULATION

Insulation: thermal, acoustic, vibration. 1957-. Bi-M. Lomax, Erskine & Co Ltd, 8 Buckingham Street, London WC2.

IRON AND STEEL

Acier/Stahl/Steel: international review for the development of the uses of steel. 1932-. M. Centre Belgo-Luxembourgeois d'Information de l'Acier, 47 Rue Montoyer, Brussels, Belgium. Editions in English, French, German, Italian and Spanish.

Archiv für das eisenhüttenwesen. 1927-. M. Verein Deutscher Eisenhüttenleute und Max-Planck-Institut für Eisenforschung. Verlag Stahleisen GmbH, Postfach 8229, Briete Str 27, Düsseldorf, Germany.

British steelmaker: a monthly review of the steel industry. 1935-. M. British Steelmaker Ltd, 7 Chesterfield Gardens, London W1.

Iron age: the national metalworking weekly. 1855-. W. Chilton Co, Chestnut & 56th Streets, Philadelphia, Pennsylvania 19139.

Iron and steel. 1927-. M. Iliffe Specialist Publications Ltd, Dorset House, Stamford Street, London SE1.

Iron and steel engineer. 1923-. M. Association of Iron and Steel Engineers, Empire Building, Pittsburgh 22, Pennsylvania.

Iron and Steel Institute. Journal. 1869-. M. Iron and Steel Institute, 4 Grosvenor Gardens, London SW1.

Iron and Steel Institute of Japan. Transactions. 1961-. M. Iron and Steel Institute of Japan, Keid duren Kaikai (3rd Floor), No 5 Otemachi I-chome, Chiyoda-ku, Tokyo, Japan. Text in English.

Stahl in English. English translation of *Stahl.* 1959-. M. Iron and Steel Institute, 4 Grosvenor Gardens, London SW1. Cover-to-cover translation.

Steel: metalworking weekly. Penton Publishing Co, Penton Building, Cleveland, Ohio 44113. Controlled circulation.

Steel facts. 1934-. BiM. American Iron and Steel Institute, 150 East 42nd Street, New York, NY 10017.

Steel international. 1965-. M. Intertech Press Ltd, 19-23 Ludgate Hill, London EC4.

Steel review. 1956-. Q. British Iron and Steel Federation, Steel House, Tothill Street, London SW1.

Steel times. 1866-. W. Industrial Newspapers Ltd, John Adam Street, Adelphi, London WC2.

Western machinery and steel world. 1921-. M. Western Machinery and Steel World Co, 43 Cleveland Street, San Francisco, California 94103.

House journals

Aciers speciaux. Q. Chambre Syndicale des Producteurs d'Aciers Fins et Speciaux, 12 Rue de Madrid, Paris 8ᵉ.

Bradley's magazine: featuring technical articles and news of developments in the metallurgical world. Irreg. Bradley & Foster Ltd, Darlaston Ironworks, Darlaston, Staffordshire.

Colvilles magazine. 1920-. Q. Colvilles Ltd, 195 West George Street, Glasgow C2.

DEW-technische berichte. 1961-. Q. Deutsche Edelstahlwerke Aktiengesellschaft, Germany.

ESC review. Q. English Steel Corp Ltd, River Don Works, Sheffield, Yorkshire.

Iron worker. 1919-. Q. Lynchburg Foundry Co, Division of Woodward Iron Co, PO Drawer 411, Lynchburg, Virginia.

John Summers review. Q. Scottish & Northwest Group, British Steel Corporation, St Ermins, Caxton Street, London SW1.

Stainless steel. 1966-. Q. Stainless Steel Development Association, 65 Vincent Square, London SW1.

Steel horizons. 1938-. Q. Allegheny Ludlum Steel Corp, Oliver Building, Pittsburgh, Pennsylvania 15222.

Steelways. 1945-. 5 per year. American Iron and Steel Institute, 150 East 42nd Street, New York, NY 10017.

US Steel news. 1936-. 8 per year. United States Steel Co, 525 William Penn Place, Pittsburgh, Pennsylvania 15219.

LEAD
House journal
Lead. 1930-. Bi-M. Lead Industries Association Inc, 292 Madison Avenue, New York, NY 10017.

LOCOMOTIVE ENGINEERING
American Railway Engineering Association. Bulletin. 1900-. 7 per year. The Association, 59 East Van Buren Street, Chicago, Illinois 60605.

Glasers annalen ZEV: zeitschrift für eisenbahnwesen und verkehrstechnik. 1877-. M. Georg Siemens Verlagsbuchhandlung, Lützowstrasse 6, I Berlin 30.

Institution of Locomotive Engineers. Journal. 1911-. Bi-M. The Institution, Locomotive House, Buckingham Gate, London SW1.

International Railway Congress Association. Monthly bulletin. 1923-. M. The Association, 19 Rue du Beau-Site, Brussels 5, Belgium.

Railway gazette: a journal of management, engineering and operation. 1902-. Semi-M. Tothill Press Ltd, 161-166 Fleet Street, London EC4.

Railway locomotives and cars. 1832. M. Simmons-Boardman. Publishing Corp, 30 Church Street, New York, NY 10007.

Railway magazine. 1897-. M. Tothill Press Ltd, 161-166 Fleet Street, London EC4.

Railway Technical Research Institute (JNR). Quarterly report. 1960-. Q. Japanese National Railways, Kunitachi Box 9, Tokyo, Japan.

Revue générale de chemins de fer. 1878-. M. Société Nationale des Chemins de Fer Français, Dunod SA, 92 Rue Bonaparte, Paris 6e.

LUBRICATION see TRIBOLOGY

MACHINE DESIGN
Journal of mechanisms. 1966-. Q. Pergamon Press, Headington Hill Hall, Oxford.

Machine design. 1929-. Semi-M. Penton Publishing Co, Penton Building, Cleveland, Ohio 44113.

Machine design and control: a technical monthly for designers of industrial and manufacturing machinery. 1962-. M. Mercury House, Waterloo Road, London SE1. Controlled circulation.

Machine design engineering. 1963-. M. Design Engineering Publications Ltd, Hermes House, 89 Blackfriars Road, London SE1.

See also ENGINE DESIGN AND APPLICATION; ENGINEERING DESIGN; FLUID POWER AND PNEUMATICS

MACHINE TOOLS see WORKSHOP TECHNOLOGY

MACHINING see WORKSHOP TECHNOLOGY

MAGNESIUM
House journals
Magplate on press. 1959-. Q. Brooks and Perkins, 1950 West Fort Street, Detroit, Michigan 48216.

Metalscope. 1947-. Q. Brooks & Perkins, 1950 West Fort Street, Detroit, Michigan 48216.

MAINTENANCE ENGINEERING see PLANT ENGINEERING AND INDUSTRIAL EQUIPMENT

MANUFACTURING see PRODUCTION ENGINEERING; WORKSHOP TECHNOLOGY

MARINE ENGINEERING
European shipbuilding: journal of the ship technical society. 1951-. Bi-M. Selvigs Forlag, Radhusgt 8, Box 162, Oslo, Norway. Text in English.

Hansa: zeitschrift für schiffahrt, schiffbau, hafen. 1864-. Semi-M. C Schroeder & Co, Stubbenhuk 10, Hamburg II, Germany.

Holland shipbuilding. 1951-. M. National Nautical and Aeronautical Institute, 10 Burgemeister s ' Jacobplein, Rotterdam, Netherlands.

Institute of Marine Engineers. Transactions. 1889-. Institute of Marine Engineers, Memorial Building, 76 Mark Lane, London EC3.

Institution of Engineers and Shipbuilders in Scotland. Transactions. 1857-. 7 per year. The Institution, 39 Elmbank Crescent, Glasgow 2.

International shipbuilding progress: shipbuilding and marine engineering monthly. 1954-. M. International Periodical Press, 194 Heemraadssingel, Rotterdam, Netherlands.

Japan shipbuilding and marine engineering. 1966-. Bi-M. Japan Association for Technical Information, Iwamoto Building, 4-12 2-chome, Misaki-cho, Chiyoda-ku, Tokyo. Text in English.

Journal of hydronautics. 1967-. Q. American Institute of Aeronautics and Astronautics, 1290 Avenue of the Americas, New York, NY 10019.

Journal of ship research. 1957-. Q. Society of Naval Architects and Marine Engineers, 74 Trinity Place, New York, NY 10006.

Marine engineer and naval architect. 1879-. M. Whitehall Technical Press Ltd, Wrotham Place, Wrotham, Sevenoaks, Kent.

Marine engineering/Log. 1878-. M. Simmons-Boardman Publishing Corp, 30 Church Street, New York, NY 10007.

Marine technology. 1964-. Q. Society of Naval Architects and Marine Engineers, 74 Trinity Place, New York, NY 10006.

Motor ship. 1920-. M. Temple Industrial Publications Ltd, Bowling Green Lane, London EC1.

Naval engineer's journal. 1880-. BiM. American Society of Naval Engineers, 1012 14th Street NW, Washington, DC 20005.

North-East Coast Institution of Engineers and Shipbuilders. Transactions. 1884-. 8 per year. The Institution, Bolbec Hall, Westgate Road, Newcastle-upon-Tyne 1.

Reed's marine equipment news. 1957-. M. Thomas Reed Publications Ltd, 39 St Andrew's Hill, London EC4.

Royal Institution of Naval Architects. Quarterly transactions. 1860-. Q. The Institution, 10 Upper Belgrave Street, London SW1.

Schiffstechnik: forschungshefte für schiffbau und schiffsmaschinenbau 1952-. 5 per year. C. Schroeder & Co, Stubbenhuk 10, Hamburg II, Germany.

Shipbuilding and shipping Record, 1913-. W. Shipbuilding & Shipping Record, 161-166 Fleet Street, London EC4.

Shipping world and shipbuilder. 1883-. W. Benn Brothers Ltd, 7-17 Jewry Street, London EC3.

House journals
  Burntisland Shipbuilding Group journal. 1920-. 2 per year. Burntisland Shipbuilding, Burntisland, Fife, Scotland.
  Crossley chronicles. Irreg. Crossley Brothers Ltd, Openshaw, Manchester 11, England.

MASS PRODUCTION see PRODUCTION ENGINEERING; WORKSHOP TECHNOLOGY

MATERIALS HANDLING
  Deutsche hebe- und foerdertechnik. 1954-. M. AGT-Verlag Georg Thum, Geotheplatz 8, Postfach 319, 714 Ludwigsburg, Germany.
  Fördern und heben: independent periodical for rationalisation and automation in mechanical handling and storing. 1951-. 22 per year. Krausskopf Verlag für Wirtschaft GmbH, Lessingstr 12-14, Mainz, Germany. Text in English, French and German.
  Handling & shipping. 1960-. M. Industrial Publishing Co, 812 Huron Road, Cleveland, Ohio 44115.
  Industrial handling. 1963-. M. Tothill Press Ltd, 161-166 Fleet Street, London EC4. Controlled circulation.
  Industrial trucks. 1958-. M. Wheatland Journals Ltd, Stamford House, 65-66 Turnmill Street, London EC1.
  Manutention—stockage: les methods modernes de levage, de manutention, de stockage. 1952-. M. Compaigne Française d'Editions, 40 Rue du Colisée, Paris 8°.
  Material handling engineering: magazine of management and flow of materials. 1945-. M. Industrial Publishing Corp, 812 Huron Road, Cleveland, Ohio 44115.
  Materials handling news. 1955-. M. Iliffe Industrial Publications Ltd, Dorset House, Stamford Street, London SE1. Controlled circulation.
  Mechanical handling. 1891-. M. Iliffe Industrial Publications Ltd, Dorset House, Stamford Street, London SE1.
  Modern materials handling. 1946-. M. Cahners Publishing Co Inc, 221 Columbus Avenue, Boston 16, Mass.
  Storage, handling and equipment news. 1967-. M. Office Publications Ltd, Mercury House, Waterloo Road, London SE1. Controlled circulation.

Storage handling distribution: international journal of materials management. 1953-. M. Wheatland Journals Ltd, Stamford House, 65-66 Turnmill Street, London EC1.
House journals
 Acrow digest. 1963-. Q. Acrow (Engineers) Ltd, South Wharf Road, London WC2.
 Conveyancer quarterly news. Q. Conveyancer Fork Trucks Ltd, Liverpool Road, Warrington, Lancashire.
 Dexion news. Q. Dexion Ltd, Dexion House, Empire Way, Wembley, Middlesex.
 Industrial review. 1961-. Q. Industrial Machinery Co, Box 1259, Fort Worth 1, Texas.
 Link-belt news. 1934-. 8 per year. Link-Belt Co, Prudential Plaza, Chicago, Illinois 60601.
 Rapid handler. 1950-. 8 per year. Rapids-Standard Co Inc, Grand Rapids, Michigan 49505.
 Space. 1949-. 2 per year. Hyster Co, 2902 NE Clackamas Street, Portland 8, Oregon.

MATERIALS SCIENCE AND ENGINEERING
 Engineering fracture mechanics. 1968-. Q. Pergamon Press, Headington Hill Hall, Oxford.
 Experimental mechanics: journal of the Society for Experimental Stress Analysis. 1961-. M. The Society, 20th and Northampton Streets, Easton, Pennsylvania 18042.
 International journal of fracture mechanics. 1965-. Q. P Noordhoff Ltd, Scientific Publications Department, Box 39, Groningen, Netherlands. Text in English, French or German.
 Journal of composite materials. 1967-. Q. Technomic Publishing Co, 750 Summer Street, Stamford, Connecticut 06902.
 Journal of materials. 1966-. Q. American Society for Testing and Materials, 1916 Race Street, Philadelphia, Pennsylvania 19103.
 Journal of materials science. 1965-. M. Chapman and Hall Ltd, 11 New Fetter Lane, London EC4.
 Journal of strain analysis. 1965-. Q. Joint British Committee for Stress Analysis. Institution of Mechanical Engineers, 1 Birdcage Walk, London SW1.

140

Materialprüfung/Materials testing/Materiaux essais et recherches.
1959-. M. vdi-Verlag GmbH, Bongardstr 3, 4 Düsseldorf 10, Germany. Text in English, French and German.

Materials engineering. 1954-. M. Reinhold Publishing Co, 430 Park Avenue, New York, NY 10022.

Materials research and standards. 1921-. M. American Society for Testing and Materials, 1916 Race Street, Philadelphia, Pennsylvania 19103.

Materials science and engineering: an international journal. 1965-. BiM. Elsevier Publishing Co, Box 211, Amsterdam, Netherlands. Text in English, French and German.

SESA proceedings. 1943-. 2 per year. Society for Experimental Stress Analysis, 21 Bridge Square, Westport, Connecticut 06880.

Soviet materials science. English translation of *Fiziko-khimicheskaia mekhanika materialov.* 1965-. 6 per year. Faraday Press Inc, 84 Fifth Avenue, New York, NY 10011. Cover-to-cover translation.

Strain. 1965-. Q. British Society for Strain Measurement, 20 Peel Street, London w8.

Test engineering and management. 1948-. M. Mattingley Publishing Co, 61 Monmouth Road, Oakhurst, New Jersey 07755. Controlled circulation.

Werkestoffe. 1959-. M. Holz Verlag GmbH, Postfach 10, D8905 Mering, Germany.

See also ENVIRONMENTAL ENGINEERING; NON-DESTRUCTIVE TESTING

MECHANICAL HANDLING see MATERIALS HANDLING

METAL FINISHING

Electrochemical society. Journal. 1948-. M. Electrochemical Society Inc, 215 Canal Street, Manchester, New Hampshire.

Electrochimica acta. 1957-. M. Pergamon Press Ltd, Headington Hill Hall, Oxford.

Electroplating and metal finishing. 1947-. M. 85 Udney Park Road, Teddington, Middlesex.

Fachberichte für oberflächentechnik. 1963-. Bi-M. L A Klepzig Verlag, 4 Düsseldorf I, Friedrichstrasse 112, Germany.

Fe+Zn. European General Galvanizers' Association. 1960-. Irreg. Zinc Development Association, 34 Berkeley Square, London WI.

Finish: internationale literaturschau für die oberflächenbehandlung von metallen. 1965-. M. Eugen G Leuze Verlag, Postfach 8, 7968 Saulgau (Württ), Germany.

Galvano: traitments et finitions des surfaces. 1932-. M. Société Galvano, 79 Champs Elysees, Paris 6ᵉ.

Galvanotechnik. 1902-. M. Eugene G Leuze Verlag, Postfach 8, 7968 Saulgau (Württ), Germany.

Industrial finishing. 1949-. M. Arrow Press, Stamford House, 65-66 Turnmill Street, London ECI.

Institute of Metal Finishing. Transactions. 1951-. 5 per year. The Institute, 178 Groswell Road, London ECI.

Journal du four électrique: électrochemie, électrométallurgie, électrothermie. 1895-. M. Publications Minières et Métallurgiques, 86 Rue Cardinet, Paris 17ᵉ.

Metal finishing: devoted exclusively to metallic surface treatments. 1903-. M. Metals and Plastics Publications Inc, 99 Kinderkamack Road, Westwood, New Jersey 09675.

Metal finishing journal: science and practice of all surface coating processes. 1930-. M. Fuel and Metallurgical Journals Ltd, John Adam House, 17-19 John Adam Street, Adelphi, London WC2.

Metal treating. 1950-. Bi-M. Temple Publications, Hanna Building, Cleveland, Ohio 44115.

Metalloberfläche: zeitschrift für galvanotechnik und alle anderen gebiete des schutzes und der veredlung metallischer oberflächen. 1946-. M. Deutscher Gesellschaft für Galvanotechnik, Carl Hanser Zeitschriftenverlag GmbH, Kolbergerstr 22, 8000 Munich 27, Germany.

Plating. 1910-. M. American Electroplaters' Society Inc, 443 Broad Street, Newark, New Jersey 07102.

Product finishing. 1948-. M. Sawell Publications Ltd, 4 Ludgate Circus, London EC4.

Products finishing. 1936-. M. Gardener Publications Inc, 431 Main Street, Cincinnati, Ohio 45202.

House journals

Finishing facts. 1956-. 2 per year. Imperial Chemical Industries Ltd, Paints Division, Wexham, Slough, Buckinghamshire.

Flambeau. Q. WPR Ltd, Rex House, Hampshire Road West, Hamworth, Middlesex.

Wild-Barfield heat-treatment. Q. Wild-Barfield Electric Furnaces Ltd, Elecfern Works, Otterspool Way, Watford Bypass, Watford, Hertfordshire.

See also PRODUCTION ENGINEERING; WORKSHOP TECHNOLOGY

METAL WORKING see WORKSHOP TECHNOLOGY

METALLURGY

ATB métallurgie. 1960-. Q. Association des Ingénieurs de la Faculté Polytechnique de Mons, 9 Rue de Houdain, Mons, Belgium.

Acta metallurgica: an international journal for the science of materials. 1953-. M. Pergamon Press Ltd, Headington Hill Hall, Oxford.

Alloy digest. 1952-. M. Engineering Alloy Digest Inc, 356N Mountain Avenue, Upper Mount Clair, New Jersey.

American metal market. 1882-. D (Mon-Fri). American Metal Market Co, 525 West 42nd Street, New York, NY 10036.

American Society for Metals. Transactions quarterly. 1961-. Q. American Society for Metals, Metals Park, Ohio.

Australian Institute of Metals. Journal. 1956-. Q. The Institute, 23 MacKillop Street, Melbourne C1, Australia.

Canadian metallurgical quarterly. 1962-. Q. Canadian Institute of Mining and Metallurgy, 1117 St Catherine Street West, Montreal, Quebec, Canada.

Indian Institute of Metals. Transactions. 1946-. Q. Indian Institute of Metals, 31 Chowringhee Road, Calcutta 16, India. Text in English.

Institute of Metals. Journal. 1909-. M. Metals & Metallurgy Trust of the Institute of Metals and the Institution of Metallurgists, 17 Belgrave Square, London SW1.

Japan Institute of Metals. Transactions. 1960-. Q. 165 3-chome, Omachi, Sendai, Japan. Text in English.

Journal of metals. 1949-. M. American Institute of Mining, Metallurgical and Petroleum Engineers, 345 East 47th Street, New York, NY 10017.

Memoires scientifiques de la revue de métallurgie. 1904- M. 24 Rue de Clichy, Paris 9ᵉ

Metal progress. 1920-. M. American Society for Metals, Metals Park, Ohio.

Metal science and heat treatment. English translation of *Metallovedenie i termicheskaia obrabotka metallov.* 1959-. M. Plenum Corporation, 227 West 17 Street, New York, NY 10011. Cover-to-cover translation.

Metal science journal. 1967-. Bi-M. Metals and Metallurgy Trust, Institute of Metals and the Institution of Metallurgists, 17 Belgrave Square, London SW1.

Metall: internationale zeitschrift fur technik und wirtschaft. 1947-. M. Metall-Verlag GmbH, 18 Hubertusallee, I Berlin 33, Germany.

Metallgesellschaft aktiengellschaft. 1929-. Metallgesellschaft AG, Reuterweg 14, Frankfurt/Main, Germany. Text in English, French and German.

Metallurgia: the international journal of metals and materials. 1929-. M. Kennedy Press Ltd, 31 King Street West, Manchester 3.

Métallurgie et la construction mécanique. 1868-. M. J Bourdel, 79 Champs Elysées, Paris 8ᵉ.

Metallurgist. English translation of *Metallurg.* 1959-. M. Plenum Publishing Corporation, 227 West 17th Street, New York, NY 10011. Cover-to-cover translation.

Metals and materials and metallurgical reviews. 1967-. M. Metals and Metallurgy Trust, Institute of Metals and the Institution of Metallurgists, 17 Belgrave Square, London SW1.

Metals engineering quarterly. 1961-. Q. American Society for Metals, Metals Park, Ohio.

Metaux: corrosion-industries. 1925-. M. Editions Metaux, 32 Rue du Maréchal-Joffre, Saint Germain-en-Laye, France.

Neue hütte: wissenschaftliche zeitschrift für das hüttenwesen und verwandte gebiete. 1956-. M. VEB Deutsche Verlag für Grundstoffindustrie, Karl-Heine-Str 27, 7031 Leipzig, Germany.

Revue de métallurgie. 1904-. M. 25 Rue de Clichy, Paris 9ᵉ.

Revue romaine des sciences techniques série de métallurgie. 1956-. 2 per year. Editions de l'Academie RPR, Str Foisorului 41, Bucharest, Roumania. Text in English, French, German, Russian and Spanish.

Russian metallurgy. English translation of *Metally*. 1962-. 6 per year. Scientific Information Consultants Ltd., 661 Finchley Road, London NW2. Cover-to-cover translation.

Zeitschrift für erzbergbau und metallhüttenwesen. 1948-. M. Gesellschaft Deutscher Metallhütten-und Bergleute ev, Dr Riederer-Verlag GmbH, Mörikestr 17, Stuttgart S, Germany.

Zeitschrift für metallkunde. 1911-. M. Gesellschaft Deutscher für Metallkunde ev, Dr Riederer-Verlag GmbH, Mörikestr 17, Stuttgart S, Germany.

House journals

Alloy metals revue. 1936. Q. High Speed Steel Alloys Ltd, Widnes, Lancashire.

AMAX journal, 1963. Q. American Metal Climax Inc, 1270 Avenue of the Americas, New York, NY 10020.

BHP technical bulletin. 1957-. 3 per year. Broken Hill Proprietary Co Ltd, 500 Bourke Street, Melbourne C1, Victoria, Australia.

Close up. 1956-. Q. Armstrong Whitworth (Metal Industries) Ltd, Closeworks, Gateshead 8, Durham.

Haynes' alloy digest. Q. Union Carbide UK Ltd, Super Alloys Dept, Shepley Street, Glossop, Derbyshire.

Technische mitteilungen Krupp. Forschungsberichte. 1943-. Irreg. Zentralinstitut für Forschung und Entwicklung, Fachbucherei, Postfach 917, 43 Essen 1, Germany.

Wellman magazine. Q. Wellman Engineering Corporation Ltd, Parnell House, Witton Road, London SW1.

See also FOUNDRY PRACTICE; METAL FINISHING; NON-FERROUS METALS; POWDER METALLURGY; WORKSHOP TECHNOLOGY.

METROLOGY

Measuring tools: metrology applied to manufacturing. M. Industrial Reviews Ltd, 109 High Street, Winchester, Hampshire.

Metrologia: international journal of scientific metrology. 1965-. Q. International Committee of Weights and Measures, Springer-Verlag, Heidelberger Platz 3, Berlin 31, Germany.

Revue de métrologie. 1923-. M. 102 Rue de la Tour, Paris 16ᵉ.

House journal

DISA information: electronic measurements of mechanical events.

1965-. Irreg. DISA (Dansk Industri Syndikat A/S), 116 College Road, Harrow, Middlesex.

See also AUTOMATION AND INSTRUMENTATION.

NICKEL

House journals

Inco nickel: an engineering review. 3 per year. International Nickel Ltd, Thames House, Millbank, London SW1.

International nickel. 1956-. 3 per year. International Nickel Ltd, Thames House, Millbank, London SW1.

Nickel topics: published in the interests of producers and users of nickel and nickel alloys. 1948-. 8 per year. International Nickel Co Inc, 67 Wall Street, New York, NY 10005.

Wiggin nickel alloys. 3 per year. Henry Wiggin & Co Ltd, Hereford, England.

NON-DESTRUCTIVE TESTING

British journal of non-destructive testing. 1959-. Q. Non-Destructive Testing Society of Great Britain, 10 Chalfont Close, Leigh-on-Sea, Essex.

Materials evaluation. 1942-. M. Society for Non-Destructive Testing, 914 Chicago Avenue, Evanston, Illinois 60202.

Soviet defectoscopy: a journal of non-destructive testing. English translation of *Defektoskopiia.* 1965-. Bi-M. Plenum Publishing Corporation, 227 West 17th Street, New York NY 10111. Cover-to-cover translation.

Testing instruments and control: official journal of the Non-Destructive Testing Association of Australia. 1963-. M. The Association, PO Box 250, North Sydney 2060, Australia.

NON-FERROUS METALS

Journal of the less-common metals: an international journal of chemistry and metallurgy. 1959-. M. Elsevier Publishing Co, Box 211, Amsterdam, Netherland. Text in English, French or German.

Light metal age. 1943-. Bi-M. 693 Mission Street, San Francisco, California 94105.

Modern metals. 1945-. M. W B Griffin, 435 North Michigan Avenue, Chicago, Illinois 60611.

Soviet journal of non-ferrous metals. English translation of *Tsvetnye metally.* Primary Sources, 11 Bleecker Street, New York, NY 10012. Cover-to-cover translation.

House journal

Metal products news. 1950-. Bi-M. Dow Chemical Co, Metal Products Department, Midland, Michigan 48640.

See also ALUMINIUM; COBALT; COPPER; LEAD; MAGNESIUM; NICKEL; PLATINUM; TIN; TITANIUM; ZINC.

NUCLEAR ENGINEERING

American nuclear society. Transactions. 2 per year. The Society, 244 East Ogden Avenue, Hinsdale, Illinois 60521.

Atom. 1956-. M. United Kingdom Atomic Energy Authority, Public Relations Branch, 11 Charles II Street, London SW1.

Atompraxis: internationale zeitschrift für wissenschaft und technik. 1955-. M. Verlag G Braun, 75 Karlsruhe 1, Postfach 1709, Bonn, Germany.

Atomwirtschaft-atomtechnik. 1955-. M. Verlag Handelsblatt GmbH, 4 Düsseldorf 1, Handelsblatthaus, Krenzstr 21, Postfach 1102, Germany.

British nuclear energy society journal. 1962-. Q. The Society, 1-7 Great George Street, London SW1.

CERN courier. 1960-. M. Publishing Office, CERN, 1211 Geneva 23, Switzerland.

Energie nucleaire: review de physique et de chemie nucleaires et de génie atomique. 1948-. Bi-M. Les Publications Techniques Associées, 29 Rue du Mon-Thabor, Paris Ier.

Euratom: review of the European atomic energy community. 1962-. Q. Euratom, Dissemination of Information Directorate, 51-55 Rue Belliard, Brussels, Belgium.

Journal of nuclear energy. 1967-. M. Pergamon Press Ltd, Headington Hill Hall, Oxford.

Journal of nuclear materials: a journal on metallurgy, ceramics and solid state physics in the nuclear energy industry. 1959-. Q. North Holland Publishing Co, PO Box 103, Amsterdam, Netherlands.

Kerntechnik: zeitschrift für ingenieure aller fachrichtungen. 1950-.

M. Verlag Karl Thiemig KG, D-8 Munchen 90, Pilgersheimerstrasse 38, Germany. Text in English and German.

Nuclear applications: a journal of applied nuclear sciences, nuclear engineering and related technology. 1965-. M. American Nuclear Society Inc, 244 East Ogden Avenue, Hinsdale, Illinois 60521.

Nuclear energy. 1947-. 6 per year. Institution of Nuclear Engineers, 147 Victoria Street, London SW1.

Nuclear engineering and design: a journal devoted to the thermal, mechanical and structural problems of nuclear energy. Bi-M. North Holland Publishing Co, Box 103, Amsterdam, Netherlands.

Nuclear engineering international. 1956-. M. Heywood-Temple Publications Ltd, 33-39 Bowling Green Lane, London EC1.

Nuclear instruments and methods: a journal on accelerators, instrumentation and techniques in nuclear physics. 1956-. M. North Holland Publishing Co, Box 3489, Amsterdam, Netherlands.

Nuclear science and engineering: a journal of the American Nuclear Society. M. The Society, 244 East Ogden Avenue, Hinsdale, Illinois 60521.

Nucleonics. 1947-. M. McGraw-Hill Inc, 330 West 42nd Street, New York, NY 10036.

Nucleonics. English translation of *Nukleonika*. 1965-. M. Clearinghouse for Federal Scientific and Technical Information, Port Royal and Braddock Roads, Springfield, Virginia.

Power reactor technology and nuclear fuel processing: a quarterly technical progress review. 1957-. Q. US Atomic Energy Commission, Superintendent of Documents, Washington, DC 20402.

Soviet atomic energy. English translation of *Atomnaya energiya*. 1956-. M. Consultants Bureau, 227 West 17th Street, New York, NY 10011. Cover-to-cover translation.

PIPING

Pipes & pipelines international: pipes, piping, hoses, tubes and auxiliary equipment. 1958-. Q. Scientific Surveys Ltd, 11A Gloucester Road, London SW7.

Röhre, röhrleitungsbau, röhrleitungstransport. 1962-. Bi-M. Verlag für angewandte Wissenschaften GmbH, Hardstr I, Baden-Baden, Germany.

House journal
Cast iron pipe news. Bi-M. Cast Iron Pipe Research Assocation, Prudential Plaza, Chicago, Illinois 60601.
See also WORKSHOP TECHNOLOGY.

PLANT ENGINEERING AND INDUSTRIAL EQUIPMENT
AIPE newsletter. 1961-. M. American Institute of Plant Engineers, Select Publishing Co, 900 Northstar Center, Minneapolis, Minnesota 55402.

Canadian industrial equipment news: reader service on new, improved and redesigned industrial equipment. 1940-. Semi-M. National Business Publications Ltd, Gardenvale, Quebec, Canada.

Engineering trader: for buyers of engineering plant and machinery. 1923-. M. Engineering Trader, Stamford House, 66 Turnmill Street, London EC1.

Factory and plant. 1963-. M. Tait Publishing Co Pty Ltd, 415 Bourke Street, Melbourne, Victoria, Australia.

Factory equipment news. 1949-. M. Production Publications (London) Ltd, 10-16 Elm Street, London WC1. Controlled circulation.

Industrial equipment news: what's new in equipment, parts, materials and literature and catalogs. 1933-. M. Thomas Publishing Co, 461 Eighth Avenue, New York, NY 1001. Controlled circulation.

Industrial intelligence. 1962-. M. Envoy Journals Ltd, 67 Clerkenwell Road, London EC1. Controlled circulation.

Industrial maintenance and plant operation. 1939-. M. Ames Publishing Co, 1 West Olney Avenue, Philadelphia, Pennsylvania 19120.

Maintenance engineering. 1957-. M. Factory Publications Ltd, 89 Blackfriars Road, London SE1. Controlled circulation.

Modern plant operation and maintenance. 1959-. Q. US Industrial Publications Inc, 209 Dunn Avenue, Stamford, Connecticut 06905.

New equipment digest: equipment, materials, processes, designs, applications, literature. 1936-. M. Penton Publishing Co, Penton Building, Cleveland, Ohio 44113. Controlled circulation.

New equipment news. 1940-. M. Canadian Engineering Publications Ltd, 46 St Clair Avenue East, Toronto 7, Canada.

Pacific factory: the plant management and production magazine of the West. 1910-. M. 693 Mission Street, San Francisco, California 94105.

Plant administration and engineering. 1941-. M. MacLean-Hunter Co Ltd, 481 University Avenue, Toronto 2, Canada.

Plant engineer. 1947-. M. Institution of Plant Engineers, Batiste Publications Ltd, Drummond House, 203-209 Gower Street, London NW1.

Plant engineering. 1947-. M. Technical Publishing Co, 308 East James Street, Barrington, Illinois 60010.

Product design and development: the news service on components, parts and materials, research and design equipment. 1946-. M. Chilton Co, Philadelphia, Pennsylvania. Controlled circulation.

PLASTICS

Applied plastics: the practical uses of plastics in industry and commerce. 1958-. M. Scientific Press Ltd, 11A Gloucester Road, London SW7.

Australian plastics and rubber journal. 1945-. Australian Trade Publications, 28 Chippen Street, Chippendale, New South Wales, Australia.

British plastics. 1929-. M. Iliffe Industrial Publications Ltd, Dorset House, Stamford Street, London SE1.

Canadian plastics magazine. 1943-. M. Southam Business Publications Ltd, 1450 Don Mills Road, Don Mills, Ontario, Canada.

Kunststoffe. English edition 1958-. M. Carl Hanser Zeitschriftenverlag GmbH, Kolbergerstr 22, 8000 Munich 27, Germany.

Modern plastics. 1925-. M. Modern Plastics Inc, 770 Lexington Avenue, New York, NY 10021.

Plastics. 1937. M. Heywood-Temple Industrial Publications Ltd, Bowling Green Lane, London EC1.

Plastics technology: the magazine of plastics processing. 1955-. M. Bill Brothers Publications Inc, 630 Third Avenue, New York, NY 10017.

Plastics world: news of materials, equipment, applications. 1943-. M. Cahners Publishing Co, 221 Columbus Avenue, Boston, Mass 2116.

Reinforced plastics. 1956-. M. Craftsman Publications, 18 Dufferin Street, London EC1.

Soviet plastics. English translation of *Plasticheskie massy.* 1960-.

M. Rubber and Technical Press Ltd., 25 Lloyd Baker Street, London WC1. Cover-to-cover translation.

SPE journal. 1945-. M. Society of Plastics Engineers Inc, 65 Prospect Street, Stamford, Connecticut 06902.

House journals

Durez plastics news. 1935-. Bi-M. Durez Plastics Div, Hooker Chemical Corp, 6 Walck Road, N-Tonawanda, NY 14121.

ICI plastics today. 1959-. Q. Imperial Chemical Industries Ltd, Plastics Division, Welwyn Garden City, Hertfordshire.

Rohm and Haas reporter. 1943-. Rohm and Haas Co, Independence Mall West, Philadelphia, Pennsylvania 19105.

PLATINUM
House journal

Platinum metals review: quarterly survey of research on platinum metals and of developments in their application in industry. 1957-. Q. Johnson, Matthey & Co Ltd, Hatton Garden, London EC1.

PNEUMATICS see FLUID POWER AND PNEUMATICS

POWDER METALLURGY

International journal of powder metallury. 1965-. Q. American Powder Metallurgy Institute, 201 East 42nd Street, New York, NY 10017.

Metal powder report. M. Powder Metallurgy Ltd, Paramount House, 75 Uxbridge Road, London W5.

P/M parts engineering. 1958-. Q. Powder Metal Industries Federation, 201 East 42nd Street, New York, NY 10017.

Planseeberichte für pulvermetallurgie. 1952-. 3 per year. P Schwarskopf, Postfach 74, 6600 Reutte, Austria.

Powder metallurgy. 1958-, 2 per year. Powder Metallurgy Joint Group of the Iron and Steel Institute and Institute of Metals, 17 Belgrave Square, London SW1.

Soviet powder metallurgy and metal ceramics. English translation of *Poroshkovaia metallurgiia.* 1962-. M. Plenum Publishing Corporation, 229 West 17th Street, New York, NY 10011. Cover-to-cover translation.

PROCESS HEATING see INDUSTRIAL AND PROCESS HEATING

Appliance manufacturer: design, engineering, management, production, finishing, marketing, purchasing. 1953-. M. Cahners Publishing Co Inc, 5 South Wabash Avenue, Chicago, Illinois 60603.

Australian machinery and production engineering. 1946-. M. Institution of Production Engineers and Institute of Machine Tools (Australia). Australian Trade Publications, 28 Chippen Street, Chippendale, New South Wales, Australia.

CIRP annals: research journal of the International Institution for Production Engineering. Q. Pergamon Press Ltd, Headington Hill Hall, Oxford.

Factory management. 1934-. M. Production Publications (London) Ltd, Elm House, 10-16 Elm Street, London WC1. Controlled circulation.

Fertigungstechnik und betrieb: zeitschrift für technologie und organisation. 1951-. M. Verlag Technik, Berlin, Germany.

International journal of production research. 1961-. Q. Institution of Production Engineers, 10 Chesterfield Street, London W1.

Klepzig fachberichte für die führungskrafte aus maschinenbau und hüttenwesen. 1882-. M. L A Klepzig Verlag, Friedrichstr 112, Düsseldorf, Germany.

Light production engineering: design, production, finishing. 1963-. M. Factory Publications, Mercury House, 103-119 Waterloo Road, London SE1. Controlled circulation.

Machinery and production engineering. 1912-. W. Machinery Publishing Co Ltd, New England Street, Brighton 1, Sussex.

Mass production. 1928-. M. Sawell Publications Ltd, 4 Ludgate Circus, London EC4.

Modern manufacturing. 1891-. M. McGraw-Hill Inc, 330 West 42nd Street, New York, NY 10036.

Product engineering. 1930. Semi-M. McGraw-Hill Inc, 330 West 42nd Street, New York, NY 10036.

Production: the magazine of manufacture. 1934-. M. Bramson Publishing Co, Box 1, Birmingham, Michigan 48012.

Production engineer. 1921-. M. Institution of Production Engineers, 10 Chesterfield Street, London, W1.

Production equipment: metalworking, machines, tools, materials, methods. 1942-. M. Wilson-Carr Inc, South Dearborn Street, Chicago, Illinois 60605.

Production equipment digest. 1954-. M. Hulton Publications Ltd, Audrey House, Ely Place, London EC1.

Werkstatt und betrieb: zeitschrift für maschinenbau und fertigung. 1868-. M. Carl Hanser Zeitschriftenverlag GmbH, Kolbergerstr 22, 8000 Munich 27 Germany.

Werkstattstechnick: zeitschrift für produktion und betrieb. 1907-. M. Springer Verlag, Heidelberger Platz 3, I Berlin I, Germany.

Works engineering and factory services. 1906-. M. Tothill Press Ltd, 161-166 Fleet Street, London EC4. Controlled circulation.

See also AUTOMATION AND INSTRUMENTATION; FABRICATION AND JOINING; FOUNDRY PRACTICE; METAL FINISHING; QUALITY CONTROL AND MEASUREMENT; WORKSHOP TECHNOLOGY.

PUMPS see FLUID POWER AND PNEUMATICS

QUALITY CONTROL AND MEASUREMENT

Quality assurance: the management magazine for applied quality control and product reliability. 1962-. M. Hitchcock Publishing Co, Wheaton, Illinois 60187. Controlled circulation.

Quality engineer. 1935-. Bi-M. Institution of Engineering Inspection, Techni-print Ltd, 45 Great Russell Street, London WC1.

Quality progress. 1944-. M. American Association for Quality Control, 161 West Wisconsin Avenue, Milwaukee, Wisconsin 53203.

House journal

Flow line: magazine of measurement and control. 1942-. Bi-M Rockwell Manufacturing Co, 400 North Lexington Avenue, Pittsburgh, Pennsylvania 15208.

See also METROLOGY

REFRIGERATION ENGINEERING

Australian refrigeration, air conditioning and heating. 1947-. M. Australian Institute of Refrigeration, Air Conditioning and Heating, Inc, Page Publications Pty Ltd, Box 606, GPO Sydney, Australia.

Cyrogenic engineering news. 1965-. M. Business Communications Inc, 2800 Euclid Avenue, Cleveland, Ohio 44115.

Cyrogenics: the international journal of low temperature engineering and research. 1960-. Bi-M. Heywood Temple Publications, 33-39 Bowling Green Lane, London EC1. Text in English, French, German, Italian and Russian.

Institut International du Froid. Bulletin. Bi-M. The Institute, 177 Boulevard Malesherbes, Paris 17e.

Institute of Refrigeration. Proceedings. Annual. The Institute, New Bridge Street House, New Bridge Street, London EC4.

Journal of refrigeration. 1957-. M. Journal of Refrigeration Ltd, 19 Harcourt Street, London W1.

Kaeltetechnik-klimatisierung: zeitschrift für das gesamte gebiet der kaelteerzeugung, kaelteanwendung und klimatisierung. 1949-. M. Deutscher Kaeltechnischer Verein, Verlag C F Müller, Rheinstr 122, Postfach 4329, 75 Karlruhe-West, Germany.

Modern refrigeration and air conditioning. 1898-. M. Davis House, 69-77 High Street, Croydon, Surrey.

Refrigeration. 1906-. M. John W Yopp Publications Inc, 1070 Spring Street, Atlanta, Georgia 30309.

Revue générale du froid et du conditionement de l'air. 1909-. M. Editions Géographiques Professionnelles, 9 Rue Coetlogon, Paris 6e.

World refrigeration and air conditioning: devoted to the refrigeration and allied industries. 1950-. M. 1 Crane Court, Fleet Street, London EC4.

House journal

Cold logic. 1951-. Q. York Shipley Ltd, North Circular Road, London NW2.

See also HEATING AND VENTILATING ENGINEERING.

SHEETMETAL WORKING see WORKSHOP TECHNOLOGY
SHIPBUILDING see MARINE ENGINEERING

STEAM ENGINEERING

Combustion: devoted to the advancement of steam plant design and operation. 1929-. M. Combustion Publishing Co Inc, 277 Park Avenue, New York, NY 10017.

Steam and heating engineer. 1931-. M. Troup Publications Ltd, 35 Red Lion Square, London WC1. Controlled circulation.

See also APPLIED HEAT; ENGINE DESIGN AND APPLICATION; MARINE ENGINEERING.

STEEL see IRON AND STEEL

STRENGTH OF MATERIALS see MATERIALS SCIENCE AND ENGINEERING

THERMODYNAMICS see APPLIED HEAT

TIN

Tin international: monthly magazine for the producers and consumers of tin. 1927-. M. Tin Publications Ltd, 7 High Road, London W4.

House journals

Notes on tin. 1956-. M. International Tin Council, Haymarket House, 28 Haymarket, London SW1.

Tin and its uses. 1939-. Q. Tin Research Institute, Fraser Road, Greenford, Middlesex.

Tin news. 1952-. M. Malayan Tin Bureau, 2000 K Street, Washington, DC 20006.

TITANIUM

House journal

Titanium progress. 1968-. Q. Imperial Metal Industries (Kynoch) Ltd, New Metals Division, PO Box 216, Birmingham 6.

TOOLING see WORKSHOP TECHNOLOGY
VACUUM TECHNIQUES see FLUID POWER AND PNEUMATICS
VENTILATING see HEATING AND VENTILATING ENGINEERING

TRIBOLOGY

ASLE transactions. 1958-. Q. American Society of Lubrication Engineers. Academic Press, III Fifth Avenue, New York, NY 10003.

Energie fluide et lubrication. 1962-. M. Union Française de Publications Techniques, 64 Rue Ampére, Paris 17e.

Friction and wear in machinery. English translation of *Trenie i iznos v mashinakh.* 1956-. M. American Society of Mechanical Engineers, 345 East 47th Street, New York, NY 10017. Cover-to-cover translation.

Industrial lubrication and tribology. 1948-. M. Scientific Publications, Broseley, Shropshire.

Journal of lubrication technology. (*Transactions of the American Society of Mechanical Engineers, Series F*) See page 116.

Lubrication engineering. 1945-. M. American Society of Lubrication Engineers, 838 Busse Highway, Park Ridge, Illinois 60068.

NLGI spokesman. 1937-. M. National Lubricating Grease Institute, 4635 Wyandotte Street, Kansas City, Missouri 64112.

Petrole-lubri-Europe. 1963-. 36 per year. Gotthardstr 44, Zurich-Thalwil, Switzerland. Text in English, French and German.

Schmierstoffe und schmierungstechnik. 8 per year. VEB Deutscher Verlag für Grungstoffindustrie, Leipzig, Germany.

Schmiertechnik und tribologie. 1954-. Bi-M. Gesellschaft für Schmiertechnik, Karl Marklein-Verlag GmbH, Benderstr 168a, 4 Düsseldorf-Gerresheim, Germany.

Tribology: lubrication, friction and wear. 1968-. M. Iliffe Science and Technology Publications Ltd, 32 High Street, Guildford, Surrey.

Wear: international journal on the science and technology of friction, lubrication, wear. 1958-. Bi-M. Elsevier Publishing Co, Box 211, Amsterdam W, Netherlands. Text in English, French and German.

House journals

Castrol review. 1946-. Q. Castrol Ltd, Castrol House, Marylebone Road, London NW1.

Esso magazine. 1949-. Q. Esso Petroleum Co Ltd, Victoria Street, London SW1.

Lubrication: a technical publication devoted to the selection and use of lubricants. 1911-. M. Texaco Inc, 117 Park Street, London W1.

Oil. 1948-. 2 per year. Burmah Oil Trading Ltd, 76 Jermyn Street, London SW1.

VIBRATION

Journal of sound and vibration. 1962-. 3 per year. British Acoustical Society, Academic Press, Berkeley Square House, Berkeley Square, London W1.

Nonlinear vibration problems. 1960-. 1-2 per year. Polish Academy of Sciences, Department of Vibrations, Swietokrzyska 21, Warsaw, Poland. Text mainly in English, occasionally in French or Russian.

WEAR see TRIBOLOGY

WELDING see FABRICATION AND JOINING

WIRE
Wire: the technical journal for the wire industry. M. E Nelles Ltd, 334 Brixton Road, London SW9.
Wire and wire products: devoted to the production of wire, rod and strip, wire and rod products and insulated wire and cable. 1926-. M. Haire Publishing Co, 299 Main Street, Stamford, Connecticut 06901.
Wire industry: international monthly journal. 1934-. M. 33 Furnival Street, London EC4.
Wire world international. 1907-. Bi-M. Michael Triltsch Verlag, Jahnstr 36, 4 Düsseldorf I, Germany.
House journal
Cambridge news. 1967-. 2 per year. Cambridge Wire Cloth Co, Cambridge, Maryland.

WORKSHOP TECHNOLOGY
Abrasive engineering. 1955-. M. Hitchcock Publishing Co, Geneva Road, Wheaton, Illinois 60187.
Abrasive methods. 1963-. M. American Society for Abrasive Methods, 330 South Wells Street, Chicago, Illinois 60606.
American machinist. 1877-. Semi-M. McGraw-Hill Inc, 330 West 42nd Street, New York, NY 10036. Controlled circulation.
Automatic machining. 1939-. M. Screw Machine Publishing Co, 65 Broad Street, Rochester, NY 14614.
Blech: fachzeitschrift für die erzeugung, veredelung und verarbeitung von band und blech, rohen und profilen. 1953-. M. Prost & Meiner Verlag, Bahnhofstr 31, Postfach 691, 8630 Coburg, Germany.
Canadian machinery and metalworking. 1905-. M. Maclean-Hunter Publishing Co Ltd, 481 University Avenue, Toronto 2, Canada.
Canadian metalworking and machine production. 1938-. M. Southam Business Publications Ltd, 1450 Don Mills Road, Don Mills, Ontario, Canada.
Cutting tool engineering. 1956-. 9 per year. Technifax Publications Inc, 410 Liberty Drive, Wheaton, Illinois 60187.

Cutting tools machines and materials. 1967-. M. Industrial Reviews Ltd, 109 High Street, Winchester, Hampshire.

Die & stamping news. 1961-. M. Compact Publications, 15936 Kinloch, Detroit, Michigan 48239. Controlled circulation.

Europäischer maschinen-markt (European machinery review). 1951-. M. Europa Fachverlag GmbH, Postfach 949, 87 Würzburg, West Germany. Text in English, French and German.

Fils, tubes, bandes et profiles. Q. Editions Ampère, 46 Rue Ampère, Paris 17ᵉ.

International journal of machine tool design and research. 1961-. Q. Pergamon Press Ltd, Headington Hill Hall, Oxford.

Japan Society of Grinding Engineers. Bulletin. 1956-. Irreg. Electro-technical Laboratory, MITI, Ginza-Higashi 8-19, Chuo-ku, Tokyo, Japan.

Machine and tool blue book: for men of action in metalworking. 1906-. M. Hitchcock Publishing Co, Hitchcock Building, Wheaton, Illinois 60187.

Machine moderne. 1906-. M. Société de Publications Mécaniques, 64 Rue Ampère, Paris 17ᵉ.

Machine shop and engineering manufacture. 1940-. M. Iliffe Specialist Publications Ltd, Dorset House, Stamford Street, London SEI. Controlled circulation.

Machinery: machines, tools and materials. 1894-. M. Industrial Press Inc, 200 Madison Avenue, New York, NY 10016.

Machinery and production engineering. W. Machinery Publishing Co Ltd, New England House, New England Street, Brighton 1, Sussex.

Machinery market and the machinery engineering material gazette. 1876-. W. Machinery Market Ltd, 146a Queen Victoria Street, London EC4.

Machines and tooling. 1959-. M. English translation of *Stanki i instrument.* Production Engineering Research Association, Melton Mowbray, Leicestershire.

Machines françaises. 1953-. Q. Société d'Edition pour la Mécanique et la Machine-Outil, 35 Rue Grande, Fontaine, St Germain en Laye, France. Text in English, French, German & Spanish.

Maschine und werkzeug: fachblatt für neukonstruktionen betriebs-

158

technik, fabrik- und werkstatt-bedarf. 1899-. Semi-M. Verlag Karl Ihl & Co, Postfach 683, 8630 Coburg Bay, Germany.

Metal forming. 1935-. M. National Forgemasters' Association, Fuel and Metallurgical Journals Ltd, John Adam House, John Adam Street, Adelphi, London wc2.

Metal products manufacturing. 1944-. M. Dana Chase Jr, York Street, Park Avenue, Elmhurst, Illinois 60126.

Metal stamping. 1967-. M. American Metal Stamping Association, 3673 Lee Road, Cleveland, Ohio 44120.

Metalworking. 1945-. M. Metalworking Publishing Co, 221 Columbus Avenue, Boston, Mass 02116.

Metalworking production. 1900-. W. McGraw-Hill Publishing Co Ltd, McGraw-Hill House, Maidenhead, Berkshire.

Metaux en feuilles: production, travail, traitements. 1965-. Bi-M. Compaignie Française d'Editions, 40 Rue du Colisée, Paris 8ᵉ.

Sheet metal industries. 1927-. M. Industrial Newspapers Ltd, 17 John Adam Street, Adelphi, London wc2.

Tool and manufacturing engineer. 1932-. M. American Society of Tool and Manufacturing Engineers, 20501 Ford Road, Dearborn, Michigan 48128.

Tooling: the journal of the gauge and tool industry. 1947-. M. Sawell Publications Ltd, 4 Ludgate Circus, London EC4.

Tooling and gaging. 1946-. M. Feder Press Inc, 13952 Meyers Road, Detroit 27, Michigan.

Tooling and production: the magazine of metalworking methods. 1934-. M. Huebner Publications Inc, 13601 Euclid Avenue, Cleveland. Ohio 44112.

TZ für praktische metallbearbeitung: technical journal for practical metalworking. 1906-. M. Technischer Verlag G Grossman GmbH, 7 Stuttgart-Vaihinger, Germany.

House journals

Argus. Q. Charles H Churchill Ltd, Coventry Road, South Yardley, Birmingham 25.

BSA: the tools group journal. 1950-. Q. BSA Tools Ltd, Mackadown Lane, Kitts Green, Birmingham 33.

Craven machine tool gazette. Bi-M. Craven Bros (Manchester) Ltd, Vauxhall Works, Reddish, Stockport, Cheshire.

Cross-hatch. Q. Micromatic-Hone Corporation, 8100 Schoolcraft Avenue, Detroit, Michigan 48238.

Grits and grinds: a technical magazine devoted to the interests of better grinding, lapping and surface finishing. 1909-. 8 per year. Norton Co, 3 New Bond Street, Worcester, Mass 01606.

Heald herald. Heald Machine Co, Worcester, Mass 01606.

Machine tool review. 1913-. Bi-M. Alfred Herbert Ltd, Box 30, Coventry.

Platt group bulletin. Q. Platt Bros (Sales) Ltd, PO Box 55, Accrington, Lancashire.

Production news. Q. Staveley Asquith Ltd, King Edward House, New Street, Birmingham 2.

Tool tips. Bi-M. Machine Tool and Cutting Tools Group, Ex-Cello-O Corporation, PO Box 386, Detroit, Michigan.

Wickman news. Q. Wickman Machine Tool Sales Ltd, PO Box 44, Tile Hill, Coventry.

See also FABRICATION AND JOINING; FOUNDRY PRACTICE; IRON AND STEEL; METAL FINISHING; PIPING; PLANT ENGINEERING; PRODUCTION ENGINEERING.

ZINC
House journal
European zinc alloy die casting bulletin. 1956-. Irreg. Zinc Development Association, 34 Berkeley Square, London W1.

CHAPTER 11

DOCUMENTS
PATENT SPECIFICATIONS; STANDARDS, DATA SHEETS; REPORT LITERATURE; CONFERENCE PROCEEDINGS; THESES; TRADE LITERATURE; GOVERNMENT PUBLICATIONS

Patent specifications are an important, if often undervalued, source of technical information. Each specification is a detailed description of a new device or method of manufacture, followed by certain claims made by the inventor. In the United Kingdom some 900 patent specifications are published by the Patent Office each week, while the comparable number of specifications published by the United States Patent Office is 1,200. The patent specification is the descriptive document which forms part of the ' letters patent ', a monopoly granted by the state for a limited period (sixteen years in the United Kingdom) to the patentee affording him the right to restrain others from using his invention for their own gain.

The patent system is not of recent origin. In Great Britain its antecedants can be traced back to the thirteenth century, to the monopolies which were granted by ruling monarchs, often as instruments of political favour. Such arbitrary dispensations of royal pleasure brought the monopoly system into disrepute as citizens who had followed a particular trade for many years suddenly found themselves debarred from their traditional livelihood when this involved the infringement of somebody's new monopoly. The need for reform led to the Statute of Monopolies Act of 1624, which imposed limitations on the powers of the Crown, and which introduced the essential criterion of 'novelty' for the granting of a monopoly. This act is the foundation stone on which all subsequent British patent law has been built. Currently patent systems are in operation in over 200 countries throughout the world. Details of the legal requirements of each system can be found in White, W W and Ravenscroft, B G: *Patents throughout the world.* New York, Trade Activities Inc, 1959, a loose-leaf reference manual kept up to date by periodical supplements.

In Great Britain the application for letters patent must be made by the inventor or by his assignee. Under British patent law the successful patent application must be:

i) Novel: a novelty search of prior British specifications of the last 50 years is made by one of the 500 British patent office examiners, each of whom has a knowledge of a particular art, to ensure that the subject of the application has not been anticipated;

ii) A method of manufacture: something that can be manufactured or which can assist in the manufacturing process;

iii) Genuinely inventive: not obvious to someone with experience in the field to which subject of the application belongs;

iv) Useful for its intended purpose.

The successful application must not be:

i) Contrary to law or morality;

ii) Contrary to natural laws;

iii) A recipe for a food or medicine.

In the United States the application must be made by the inventor himself, and under United States patent laws it is possible to patent any new, useful, unobvious method of manufacture, useful art, composition of matter or asexually produced plant.

Many thousands of patent specifications are published throughout the world each week. In addition to the United States and the United Kingdom who publish about 60,000 and 45,000 specifications per year respectively, other important patent publishing countries are: France, 40,000 specifications per year; Italy, 30,000; Japan, 28,000; Canada, 22,000; Germany (FDR), 16,000; Belgium, 15,000. It should be noted that not each one out of these thousands of specifications published internationally each week is unique; a high percentage of these documents are 'equivalent' specifications. As letters patent are only valid in the country in which they have been granted, a British company wishing to obtain protection in the United States, Japan, France, Germany etc, would need to apply for letters patent in each of these countries, so that in addition to the basic British patent the company would have American, Japanese, French, German and other equivalents. Two thirds of all British applications are also filed abroad, and one of the problems facing the technical librarian and information officer in industry is the identification of equivalent specifications.

French, Belgian, German and Dutch specifications are of particular importance as sources of technical information, as these documents are made public very soon after the application, with only 'formal' rather than 'novelty' examination. There is a delay of approximately two and a half years between the date of the British patent application and the date of publication of the specification, largely due to the novelty examination, but an equivalent French specification, for instance, can be available for public inspection within twenty months, and the equivalent Belgian specification within four months of the date of filing in these countries. Keeping a watch on French, Belgian, German and Dutch specifications can thus give a company early warning of a competitor's applications which may be under prosecution at the British patent office.

The layout of the patent specifications of most overseas countries follows that of the British specification, the initial page of which is headed by the name of the inventor and the specification number. British specifications were numbered consecutively from the year 1617 to the year 1851, *ie* numbers 1-14359. From 1852 to 1915 a new sequence commenced each year, and the present sequence commenced with specification number 100,001 in 1916 and has currently reached 1,100,000. Immediately following the name of the inventor can be found filing details of the specification (application date, date of original application in another country, date of publication of the specification, etc), the 'index at acceptance' classification or the 'press mark' which is allocated by the Patent Office examiners, and a general and usually uninformative title often commencing with the phrase 'Improvements in and relating to . . .'. The text of the specification can be divided into three parts:

i) Introductory matter: a formal plea on behalf of the applicant that letters patent be granted to him, followed by a statement of the field of technology to which the invention relates, which can sometimes include references to other attempts to solve the problems which the inventor claims to have overcome.

ii) The body of the specification, commencing with the 'consistory clause' or 'statement of invention', which is usually introduced by the phrase 'according to this invention . . .' and which describes the essential features of the invention. The consistory clause is followed by a comprehensive description of the invention and the drawings. It should be noted that under British patent law there is no obligation on the inventor to disclose *why* his invention works by explaining underlying, theoretical principles.

iii) The claim or claims, commencing with the phrase ' what I claim is . . .', complete the specification and serve to define the exact scope of the monopoly claimed by the applicant. In British specifications the most important claim will be cited first, with each subsequent claim being a qualification of the previous one. Those wishing to extract technical information from a patent specification quickly should refer immediately to either the consistory clause or its repetition in the first claim, as these will both give a concise account of the subject of the invention. Most indicative abstracts of patent specifications are merely ' depatentesed ' versions of the first claim. ' Patentese ' refers to the rather abstruse legal style in which many patent specifications are drafted.

The information disclosed in patent specifications can be used in several ways :

i) If a device or method of manufacture which an organisation wishes to use has been patented, the organisation can apply to the patentee for a licence to work the invention and this may be granted according to a scale of royalty payments.

ii) If the validity of a patent in which a company is interested is in question, the interested party can oppose the granting of the unsealed patent, or apply for a revocation of the existing patent on a number of grounds. One of the most common grounds for opposition is prior publication, where the opposing party will produce a document which has been publicly available in the United Kingdom at any time during the last fifty years which completely anticipates the subject of the patent under attack.

iii) Only the matter in the claims of the specification is covered by the grant of the monopoly. Thus any information disclosed in the body of the specification which is not covered by the narrower claims is freely available for public use.

iv) British patents are often allowed to lapse without exploitation. The patentee may be unable to find a sponsor with the capital and vision to promote his invention, or the invention might be patented before a suitable material is available for its development or before

164

the industry has developed the technical links necessary for its economic use. If a patent has lapsed before the end of its maximum term of sixteen years through the non-payment of renewal fees, the information disclosed in the specification is freely available for use. Renewal fees become payable annually for British patents from the end of the fourth year. United States patents, however, remain in force for their full term of seventeen years unless they have been revoked.

v) Ideas disclosed in patent specifications relating to one field may stimulate developments in a totally different field. An engineer working on cigarette-making machines was examining the problems of designing a mechanism which would automatically splice an exhausted reel of paper onto a full reel as soon as it had run out without stopping the machine. No similar work had been done on cigarette making machines but the printing industry had faced an analogous problem. A survey of patent specifications relating to printing machines gave the engineer the idea for the solution to his problems.

vi) Although patent specifications are published as legal documents and are not designed as examples of technical literature, taken collectively they can be considered as the most comprehensive ' textbooks ' available, in that they reveal the complete technical development of any branch of engineering by documenting the various attempts that have been made to solve particular problems. In many fields patent specifications often represent the only available source of technical information. That patent specifications can be a useful source of information for mechanical engineers can be gauged from the fact that the second largest number of current British patents fall within the field of machine elements.

The main tools for selecting currently published patent specifications and for searching retrospectively for specifications on a particular topic are the official publications of the various national patent offices. In addition, some useful non-official abstracting services are issued by commercial publishers, and notifications of newly published patent specifications are given in many technical and trade periodicals, examples being *Engineer; Heating and ventilating engineer; Foundry; Industrial lubrication; Machine design; Plastics; Plating; Welding and metal fabrication; Wire and wire products; Works engineering and factory services.* Many abstracting journals also include patent specifications within their coverage.

The chief selection guide to current British patent specifications is the weekly *Official journal (patents)* (*OJ*), which publishes lists of complete specifications accepted, six weeks before their publication date, arranged by specification number, subject classification and patentee. The main sequence under specification number provides full filing details in addition to the official title. An alphabetical list of applicants for letters patent, with each name accompanied by a very brief indication of the subject of the application and the application number, is also given in each issue of the *OJ*. Yet another useful feature of this publication is its weekly list of application numbers for which complete specifications have been published. This enables interested parties to trace the progress of applications of which the application numbers have been noted.

The *Abridgments of specifications* are the main tools for searching retrospectively for British patent specifications. The abridgments, each of which is a detailed abstract of a specification, accompanied by complete filing details and an illustration, are based on a *Classification key* and are issued each week concurrently with the specifications which they cover inside twenty five subject groups. The *Classification key* is divided into eight sections lettered A-H; each section is then divided into two or more divisions, forty in all, with each division again being divided into a number of headings, 405 in all. Section F of the classification covers mechanics, lighting and heating and is divided into four divisions:

FI : Prime movers

F2 : Machine elements

F3 : Armament

F4 Lighting, heating, cooling, drying:

FI is divided into a number of headings:

FIA Hydraulic reciprocating motors, pumps, etc

FIB Internal combustion engines

FIC Centrifugal, centripital, screw pumps, etc

FIE Injectors and ejectors

FIF Rotary engines, pumps, etc

FIG Gas turbine plant

FIJ   Jet propulsion plant
FIK   Starting, stopping, reversing engines, etc
FIL   Gas turbine etc, combustion chambers
FIM   Steam engines, etc
FIN   Compressing gases, etc
Etc.

Each of these headings is again subdivided into highly specific press marks. The *Classification key* was developed basically as an aid to Patent Office examiners in the undertaking of their prior art searches, and each specification is classified by the scheme before publication. The twenty five groups of the abridgments each comprise of one or more of the forty divisions of the *Classification key*. The abridgments are issued in sheet form to be bound up into volumes, each covering a series of 25,000 specifications, and within each volume the individual abridgments are arranged by specification number. Each published volume of the abridgments also contains the relevant sections of the *Classification key* and also a *Subject matter index* at press mark level. It should be noted that several abridgments for the same specification will appear in the various groups if the examiner has allocated more than one press mark to a specification which covers several subjects. References are given at the end of each abridgment to other abridgments for the same specification in other groups.

The *Reference index to the classification key* is arranged in three sections and is the key to retrospective searching for British specifications. The initial section is an alphabetical index of over 5,000 catchwords, with an indication under each catchword of the heading of the classification under which the subject has been subsumed. This heading will then further indicate at which division and thus under which group of the abridgments the subject has been included. The second part outlines the structure of the *Classification key,* while the final section provides scope notes to indicate which subjects have and which have not been included under the various headings.

The searcher for British specifications on, for instance, gas turbines for jet propulsion, would consult the catchword index in the *Reference index to the classification key* to find the entry:

Turbine(s)
. gas turbines

.. plant

... adapted for jet propulsion   FIJ

This would lead him to abridgment group FI. By scanning the *Classification key* which he will find bound in with the abridgment volume he can narrow his search down to a more specific press mark dealing with, for instance, particular turbine components. The specific press marks and the numbers of the specifications which have been classified at each mark are listed in the *Subject matter index* of each abridgment volume, and thus the searcher is led to the numbers of the actual specifications he requires.

The *Index to names of applicants in connection with published complete specifications* enables any searcher to ascertain which patents have been granted to an individual or organisation. Each volume covers a series of 25,000 specifications, and each entry gives name of patentee, brief title of specification and specification number.

Complete sets of British patent specifications and official indexes are available on open access at the National Reference Library for Science and Invention (formerly the Patent Office Library), and less complete sets are also available at provincial reference libraries.[1] Although it is possible to undertake a patent search at many of the provincial libraries, searching is made easier at the National Reference Library by the availability there of a number of unique indexes.[2] Patent searching can be a tedious and time-consuming task, but it is possible to simplify searching procedures by, firstly, ascertaining the press mark for the subject under search, and then applying to the Patent Office for a file list. These lists, which are produced by mechanical sorting devices to fulfil specific orders only, include all specification numbers which have been allocated to a particular press mark over a period of fifty years. Once the file list has been obtained, the searcher has simply to turn up the abstracts of the specifications in the abridgments to determine which specifications are relevant to his interests.

The US Patent Office publishes the weekly *Official gazette (OG)*, which, in addition to giving legal patent decisions and other information, includes abstracts of the US specifications issued each week. The abstracts are listed within three sections (general and mechanical; chemical; electrical) in classified order according to a scheme consisting

168

of thirty classes which are further subdivided into 60,000 sub-classes. The *OG* contains, in addition to the abstracts, a name index of patentees and a classified breakdown of accepted specifications to indicate which specifications have been published on a particular subject each week. No details of applicants for letters patent are included in the *OG*. The US Patent Office also issues an annual *Index of patents*, which contains an alphabetical list of patentees and a classified breakdown of specifications by class and sub-class. The *Classification bulletins* are also published by the Patent Office to supplement the *Manual of classification* by detailing changes in the classification of patents, and by giving definitions of new and revised classes and sub-classes. A searcher for American specifications can discover the class and sub-class number he requires by consulting the US Patent Office *Manual of classification*. The searcher for specifications on pressure relief valves would begin by consulting the *Index to classification* and would find the entry:

	Class	sub-class
Valves		
pressure relief	137	455

To discover which specifications have been issued on this subject he would then refer to the weekly issues of the *OG*, checking the general and mechanical section under Class 137-455, and finally the classified section of the annual *Index of patents*.

The public search room of the US Patent Office contains the only complete collection of US specifications arranged in class and sub-class order; in this room the sub-class bundles are available on open access. Those who are not able to visit the search room can obtain sub-class lists of specification numbers, which are produced by the US Patent Office in response to individual orders. Files of US specifications arranged by specification number and sets of the *OG* can be found in twenty two depository libraries throughout the United States.

Details of the official publications of other overseas patent offices which cover their specifications are listed in *Technical information sources*.[3]

The most significant commercial publisher of patent abstracting journals is Derwent Publications Ltd (128 Theobalds Road, London wci). Four publications issued by this company list each specification or application currently published within a particular country, *British patents abstracts* (*BPA*), *German patents gazette* (*GPG*), *German patents abstracts* (*GPA*), and *Soviet inventions illustrated* (*SII*).

BPA is published each Tuesday, the abstracts appearing four weeks after the publication of the specifications. These abstracts give full details of the main claim of each specification and are classified according to Derwent's own patent classification, but each specification is abstracted under one class only. The weekly *GPG* includes abstracts of all the 1,700 German patent applications which are now issued as *offenlegungsschriften* eighteen months from the filing of the application in Germany. *Offenlegungsschriften* are issued without a novelty examination and their early disclosure date renders them a valuable source of patent information; after a novelty examination, which is made only after a request has been filed, the applications are published as *auslegeschriften*. *GPA* is also issued weekly and includes abstracts of the *auslegeschriften* five weeks after their publication. *SII* is published monthly in three separate sections (chemical; electrical; mechanical and general) six weeks after the Russian specifications or author's certificates are received in the United Kingdom.

Derwent's weekly *Belgian patents report, French patent abstracts, Japanese patents report* and *Netherlands patents report* cover chemical specifications only; however, the Derwent classification for chemical subjects includes plastics, metal finishing and metallurgy. The *Central patents index* (*CPI*) will be available from 1970, and will initially cover the chemical (including plastics and metallurgical) specifications of thirteen countries. The *CPI* service will include alerting bulletins, basic patent abstracting journals, copies of basic specifications (in English translation in the case of Japanese and Russian specifications), cards for manual or machine retrieval and records of the specifications on magnetic tape.

Details of current Japanese patent specifications are listed in the monthly *Japan patent news* (International Patent Service (INTERPAS), PO Box 101, s-Hertogenbosch, Netherlands), which is available in seven separate sections, three of which (sections II mining, metals, chemistry; IV prime movers, electrical power, tools, etc; V transportation, etc) are relevant to the scope of this guide.

STANDARDS

Standards and specifications are documents which state how materials and products should be manufactured, measured, tested or defined: they are documents which lay down sets of conditions to be fulfilled. Standards have evolved whenever repetitive operations have been developed into organized procedures. Mass production is impossible without standardization; the prosperity of the American economy dates from the period when men like Henry Ford realized the immense market potentialities for mass produced standardized goods. Standardization avoids wastage of resources, and time and money are saved when a standard is available offering a common language, a judicious procedure and interchangeability or a guarantee of fitness for use.

The American National Standards Institute (ANSI) has defined specification as ' a concise statement of the requirements for a material, process, method, procedure or service, including whenever possible, the exact procedure by which it can be determined that the conditions are met within the tolerances specified in the statement; a specification does not have to cover specifically recurring subjects or objects of wide use, or even existing objects.' A standard is the evolution of the specification, and has been defined by the same body as ' a specification accepted by recognised authority as the most practical and appropriate solution of a recurring problem '.[4] A standard may only remain effective for a limited period before being rendered obsolete by progress in technology: a more suitable material may be developed for a specific purpose or a better method of testing may be evolved. Standards are essentially dynamic, not static, and if a better method of testing a material or carrying out an operation is achieved it will be codified into a new standard.

Standards and specifications can be conveniently divided into: dimensional standards; standards of performance or quality; standards of testing; standards of nomenclature or terminology; codes of practice.

Dimensional standards specify the dimensions needed to achieve interchangeability or to make things fit, *eg* nuts and bolts; dimensional standards for machine components facilitate the early replacement of used or damaged parts.

Standards of performance or quality will ensure that a product is

171

adequate for its intended purpose, *eg* that oil heaters conform to minimum safety requirements.

Standards of testing enable materials and products intended for the same purpose to be compared; methods have been formulated, for example for the testing of elasticity, creep, tensile strength, etc of materials.

Standardized terminology enables engineers working within a particular industry to communicate more exactly. A standardized vocabulary will ensure that a term has a hard, clear, unambiguous meaning. British standard glossaries have been published for mechanical engineering (BS 2517); iron and steel (BS 2094); sheet metal working (BS 4342); refrigeration (BS 1584), etc. The symbols used in engineering must also be standardized if confusion is to be avoided, and thus several standardized compilations of symbols and abbreviations have been published by the various national and international standardizing bodies.

Codes of practice cover the installation and maintenance of equipment, and exist for such subjects as the guarding of machines and the installation and maintenance of underfeed stokers.

Standards and specifications are issued by industrial companies; trade associations and technical societies; government departments; national standardizing bodies; international standardizing bodies.

Company standards are developed by individual companies when either a national standard or a trade standard does not exist to suit the company's specialised needs. These documents are generally not available outside the individual company.

Trade association and technical society standards. Standards are developed by trade associations and technical societies to meet the needs of the industries which they serve. The most significant technical society issuing standards is the American Society for Testing and Materials (ASTM) (1916 Race Street, Philadelphia, Pennsylvania 19103), founded in 1898 as an international, non-profitmaking technical, scientific and educational society devoted to ' the promotion of knowledge of the materials of engineering and the standardization of specifications and methods of testing '. A collection of ASTM standards, the *Book of ASTM standards* is issued annually in thirty two volumes, each volume containing specifications relating to a particular field; *eg* Part 1 covers steel piping materials; 2, ferrous castings, etc. A separate index to the complete range of ASTM standards is issued annually.

The Air Conditioning and Refrigeration Institute (1815 North Myer Drive, Arlington, Virginia 22209) is a trade association serving the entire industry which has issued standards covering all types of equipment and components.

The American Society of Heating, Refrigerating and Air Conditioning Engineers (345 East 47th Street, New York, NY 10017) publishes standards, methods of rating and testing and recommended practices. The *ASHRAE guide and data book* includes an alphabetical listing of 200 standards and codes issued by sixty organisations covering air conditioning, heating and refrigerating.

The American Society of Lubrication Engineers (838 Busse Highway, Park Ridge, Illinois 60068), whose standards activity commenced in 1962, develops and promulgates standards on industrial lubricants and lubricating practices. The standards issued to date include those covering machine tool lubricants and machine bearing lubricants.

The American Society of Mechanical Engineers (345 East 47th Street, New York, NY 10017) was a founder member of ANSI, and is the sponsor or joint sponsor for thirty five ANSI committees on standards covering the whole range of mechanical engineering. In addition to this liaison work with ANSI the society has developed the *ASME boiler and pressure vessel code, 1968*, which is divided into seven sections and which contains rules for the construction of power boilers to be used in stationary service, miniature boilers, nuclear pressure vessels, heating boilers and unified pressure vessels, in addition to the care of power boilers in service and rules for welding qualifications. The society's *Power test codes* give standard directions for conducting acceptance tests and for determining the performance of power generating and testing equipment.

The American Society of Tool Manufacturing Engineers (ASTME) (20501 Ford Road, Detroit, Michigan 48128), through its National Standard Committee, encourages and promotes critical review of existing standards and the development or proposal of new standards by its chapters. ASTME acts as a clearinghouse between its members and ANSI, and as such is represented on several ANSI committees covering subjects which include small tools and machine elements and the standardization of gears.

The American Welding Society (345 East 47th Street, New York, NY 10017) has published over sixty standards classified into four broad categories: welding fundamentals; welding processes; inspection of welds; industrial applications of welding. The society has also published the useful *Index of welding standards from 21 nations*, which lists each series of national standards in a classified sequence, giving details in English, French and the language of publication.

The Society of Automotive Engineers (485 Lexington Avenue, New York, NY 10017) publishes its standards, with the exception of its standards and material specifications for the aerospace industry, which appear in loose-leaf form, annually in the *SAE handbook*. SAE standards and recommended practices are mainly developed for automotive applications but are also widely used in other fields of engineering.

In the United Kingdom the Society of British Aerospace Companies (29 King Street, London SW1) is responsible for promoting standardization within the aeronautical engineering field, and the society publishes its standards in the multi-volumed *Standards handbook* and the two volumed *Handbook of AGS parts*.

The Society of Motor Manufacturers and Traders (SMMT) (Forbes House, Halkin Street, London SW1), established in 1955, reviews the standards adopted by the British motor industry. Some of the standards thus adopted are SMMT's own, published in *Standards for the British automotive industry*, while others adopted are national standards issued by the British Standards Institution or those issued by other trade associations or technical societies such as SAE standards.

A particularly useful set of German technical society standards are the *VDI Richtlinien (German engineering guide lines)*, issued by Verein Deutscher Ingenieur, the German Association of Engineers. The scope of these guide lines is similar to that of British standard codes of practice, but they include more detail, embracing, for instance, instructions to technical personnel on plant operation. They differ from German national standards (DIN) in that they reflect the current state of the art of a particular topic and are not written in the form of recommendations. Currently these guide lines are available for the following subjects, with the numbers issued to date in brackets: production engineering (210); power engineering (37); precision engineering (25); materials handling (145); gears (41); heating (18); ventilation (14); control engineering (30); design (8); metrology (6); vibration technology (4). Further information about this series and copies of individual standards are available in the United Kingdom from The Ministry of Technology, Abell House, John Islip Street, London SWI.

Government departments: In the United Kingdom the Ministry of Technology publishes *DTD aerospace material specifications.* These publications, like all government departmental standards, are issued ' to meet a limited requirement not covered by an existing British standard, or to serve as the basis for inspection of material, the properties and uses of which are not sufficiently developed to warrant submission to BSI for standardization '. DEF specifications are published by the Ministry of Defence, again where no suitable national or commercial specification is available to cover the standardization of stores and technical procedures used by the defence services and other government departments. The War Office Directorate of Standardization issues SDM standards (*Standardization design memoranda*) and SSM standards (*Standard stores memoranda*) for use by the service departments. SDM relate to engineering and associated practices, design criteria, etc, while SSM cover lists of equipment, stores and components.

The US Department of Defense (Office of Technical Data and Standardization Policy, Washington, DC 20301) issues military specifications and standards to cover materials, products and services required by the Department and by other federal agencies where appropriate. The Defense Standardization Program is the largest and most comprehensive standards programme in the world, governmental or non-governmental, and although the bulk of its activity is directed towards military science and combat items, many military specifications and standards are directly applicable to civilian industry, and consequently these standards are regularly used by commercial firms. An annual *Index to military specifications and standards* is available from the Superintendent of Documents, US Government Printing Office, Washington, DC 20402. The General Services Administration (Federal Supply Service, Standardization Division, Washington, DC

20405) issues federal specifications and standards to cover all materials and supplies required by civil departments of the government, and also by military agencies where needed. Federal standards are issued within five groups; supply standards (qualities, types, sizes, etc); test method standards; material standards; engineering standards; procedural standards. Again, many federal standards are applicable in private industrial situations. An annual *Index to federal specifications and standards* is available from the Superintendent of Documents at the above address.

National standardizing bodies: Most countries have established national standardizing bodies which are responsible for developing and promoting standards for application within the country. The first such body, the British Standards Institution (BSI), was founded in 1901, as the Engineering Standards Committee, by the Institution of Civil Engineers, to develop standards for application within the civil engineering industry. The committee subsequently extended its activity into the mechanical and electrical fields, was granted a Royal Charter in 1929 and changed its name to BSI in 1931. Today national standardizing bodies exist in more than fifty countries. BSI's basic function is to formulate standards and codes of practice for voluntary adoption within the United Kingdom by industry and government. Standards are prepared by 4,000 technical committees, whose 20,000 members are drawn from industry, research institutions, the academic world and consumer groups. Currently there are almost 5,000 British Standards available, and an average of 350 new standards is published annually. A British standard has no mandatory implications; it is not an instrument for enforcing government control. BSI receives an annual grant from the Board of Trade, but it is an independent body and its standards are a basis for voluntary application. In addition to members subscriptions and the sale of BSI publications, BSI's other main source of income is from the endorsement of manufactured products with its certification mark. A manufacturer can arrange for BSI to examine and test his products to ensure that they have been manufactured according to BSI requirements. The presence of the BSI ' kitemark ' indicates to consumers that the articles so marked are considered fit for their intended purpose.

Standards are invariably referred to and identified by a prefix and

176

a running number, which are assigned to the standards by their originator, thus the British standard for *Gear hobbing machines for turbines and similar drives* is referred to as BS 1498, and the German national standard for *Tapered roller journal bearings* as DIN 720. The prefixes used by the other chief national standardizing bodies throughout the world are listed in *Technical information sources.*[5]

BSI publishes its *BS yearbook,* which contains abstracts of its standards arranged by the standard number. A detailed subject index to this volume enables the enquirer to ascertain which British standards exist for a particular topic. A second volume of the *Yearbook* covers the recommendations of the various international standardizing bodies. BSI has also issued a series of *Sectional lists of standards* which simply list the titles and numbers of British standards under broad subject headings. In additional to *Mechanical engineering* the lists cover: *Aerospace; Drawing practice; Metalwork; Industrial instruments; Iron and steel; Machine tools, Non-ferrous metals; Nuclear engineering; Shipbuilding* and *Welding.* Details of new British standards and standards under development are published in the monthly *BSI news.* BSI's *Overseas and Commonwealth standards* contains lists of standards which have been received by the BSI library and which are available for loan.

The American National Standards Institute (ANSI), which has previously been known first as the American Standards Association and then the United States of America Standards Institute, was originally founded by the five major American technical societies, of which one was the American Society of Mechanical Engineers. ANSI is now an association of more than 130 technical societies and trade associations and more than 2,000 company members. As with BSI, ANSI coordinates the development and approval of voluntary national standards by harnessing the combined capabilities of industry, consumers, organised labour and government. Currently there are over 3,000 ANSI standards available for use. The annual *Catalog of American standards* lists standards by their prefixes within twenty five groups. Mechanical engineering standards are issued within group B, which is again more specifically sub-divided into further divisions, *eg* B1 screw threads; B3 ball and roller bearings; B5 machine tools and components. Standards are then numbered sequentially inside each division, *eg* B1.1 unified screw threads; BI.2 gauges and gauging for unified screw threads. Each issue of the *Catalog,* which also includes ISO recommendations, is equipped with an alphabetical subject index to indicate which standards exist for a given topic. Similar compilations are issued by the other national standardizing bodies, *eg* the *Catalogue des normes français* of the Association Français de Normalisation. Deutsches Normenausschus, the German standards body, issues a classified annual *DIN English translations of German standards* in addition to its main *Normenblatt verzeichnis,* the comprehensive classified index to German standards. BSI have also recently produced an English translation of the Russian standards catalogue.

A useful guide to standardization in the United States is the National Bureau of Standards (NBS) publication, Hartman, J E: *Directory of standardization activities* (MP 288, 1967), which provides detailed information on the work of over 400 American governmental and non-governmental organisations involved in standardization. Struglia, E J: *Standards and specifications: information sources,* Detroit, Gale Research Co, 1965, in addition to listing directories and indexes of standards issued in the United States, gives information on the standards available from American governmental and non-governmental sources.

Besides the national standardizing bodies which formulate standards for industrial and general application, some countries have established organisations to develop standards for the determination of physical constants. In the United States NBS is responsible for determining the national standards of physical measurement upon which all measurements used in the United States are based. The specialized sections of NBS, such as the Product Standards Section of the Office of Engineering Standards, cooperate extensively with ANSI and ASTM. The results of the bureau's work have been published as papers in NBS journals, *eg Journal of research: Section C: Engineering and instrumentation;* as circulars, *eg Circular C585: Measurement of thickness;* or as handbooks, *eg Handbook H28: Screw-thread standards for federal services.* Details of all NBS publications issued since 1901 are listed in *Circular 460, Miscellaneous publication 240* and *Special publication 305* and their respective supplements. The Measurement Group of the National Physical Laboratory performs similar functions

178

for the United Kingdom. The division also cooperates with BSI in the preparation of standards for engineers' measuring equipment and related apparatus. A recent joint NPL/BSI project has been the revision of BS 891 covering the Rockwell hardness test, a mechanical method of determining the hardness of materials. The results of the division's work are published sometimes as papers in the primary journals or sometimes as *NPL standard reports*. A complete listing of NPL publications is given in HMSO *Sectional list no 3: Ministry of Technology publications*. Contributions of NPL staff to journals and published conference proceedings are listed in the NPL annual report, which also includes details of *NPL papers and reports*.

International standards: The existence of divergent national standards can be a barrier to international trade; although differences in national standards for the same article or component may be slight, they may mean that a manufacturer will have to produce differing versions of his products for trade with a range of foreign countries. The International Organisation for Standardization was founded in 1946 ' to reach international agreement on industrial and commercial standards and thus facilitate international trade. . . .' ISO membership consists of over fifty national standardising organisations, and ISO's 100 technical committees have to date produced over 500 recommendations, each a framework upon which more substantial, but essentially harmonious, national standards can be built. ISO issues its annual *Memento,* which gives details of member bodies, technical committees and ISO recommendations. Other international organisations which have developed standards and recommendations in collaboration with ISO include the International Congress on Combustion Engines (10 Av Hoche, Paris 8e), the International Institute of Refrigeration (177 Boul Malesherbes, Paris 17e), and the International Institute of Welding (54 Princes Gate, Exhibition Road, London SW7).

Metallurgical specifications: Metallurgical specifications are particularly well documented. A useful card index to steel specifications to which any enquirer may refer is housed and maintained at the Sheffield Commercial and Technical Library. The *SINTO index of steel specifications* now contains well over 15,000 entries, most of which refer to descriptive trade catalogues issued by the manufacturers of the various steels. Several guidebooks to metallurgical and

179

other specifications are listed in chapter 12, under the appropriate subject with the subheading ' Handbooks and data books '. Of particular relevance here is Woldman, N E: *Engineering alloys*, New York, Reinhold, fourth edition 1962, a guide to the specifications of over 35,000 alloys manufactured in the United States and abroad; and the British Steel Castings Research Association's *British and foreign specifications for steel castings*, third edition 1968, a guide to British specifications and their foreign equivalents. This publication also includes a useful guide to code letters detailing standardising bodies encountered on steel castings specifications. Miscellaneous publication no 120, National Bureau of Standards, *Standards and specifications for metals and metal products*, contains in full or gives either an abstract or a cross-reference to over 1,600 nationally recognised standards and specifications covering ferrous and non-ferrous metals.

For standards users in the United Kingdom a copy of any overseas national standard or any international recommendation may be borrowed from the BSI library. A particularly useful service maintained by the library is its card subject index to these overseas national standards. The library in addition holds some of the most important sets of non-national standards, such as ASTM standards and American federal specifications and standards, and acts as a clearinghouse for information on British and overseas non-national standards by, for instance, putting an enquirer in touch with a library willing to supply a copy of such a standard on loan. An enquirer for American military specifications would thus be referred to the Ministry of Technology's TRC Centre at St Mary Cray, Orpington, Kent. BSI's ' Technical help to exporters ' (THE) service is operated from its Hemel Hempstead Testing Centre, and is designed to assist smaller firms in coping with the complicated foreign standards and regulations which are now in force. THE has published several series of technical reports, each series in a selected field dealing with the standards and mandatory regulations of a particular country. One series covers boilers and pressure vessels for nineteen countries, other series are planned for hoists and lifting tackle, machine tools and general machinery.

In the United States the ANSI library also serves as a clearinghouse for information on standards from every source, in addition to maintaining a loan collection of over 100,000 standards.

DATA SHEETS

The Engineering Sciences Data Unit of the Royal Aeronautical Society has been producing data sheets and data memoranda for the British aeronautical industry for more than twenty five years. These sheets aim to make available to the design engineer in a convenient form authoritative information selected by a team of specialists from a study of the published literature and also, where possible from unpublished sources. The value of these data sheets was noted in 1963 by the Fielden Committee on Engineering Design, which recommended that other professional societies should undertake similar work. Consequently the Institution of Mechanical Engineers and the Institution of Chemical Engineers have embarked upon data projects with the financial assistance of the Ministry of Technology. Three series of data sheets are now issued by the Aeronautical Sciences Data Unit in cooperation with the three professional institutions: the *Aeronautical series* (the most extensive, being issued in subseries covering aerodynamics, dynamics, fatigue, performance, etc), the *Mechanical engineering series* (available from the Institution of Mechanical Engineers) and the *Chemical engineering series*. The unit has published the *Engineering sciences data index,* 1966-, a key-word subject index to the data sheets which have been issued within the three series. *Data activities in Britain,* third edition, 1969, is published by the Office for Scientific and Technical Information (OSTI) of the Department of Education and Science, and attempts comprehensive coverage of compilations of physical data in progress and related activity. The guide, which includes sections on thermodynamic and thermophysical properties, physical and mechanical properties of materials, and design, contains over 100 entries, each of which gives information of the organisation and coverage and analysis of a data project, together with details of any related publications. News of national and international data programmes, including details of published data compilations, appear in *Codata newsletter,* which is published by the Committee on Data for Science and Technology (CODATA), Westendstrasse 19, 6 Frankfurt/ Main, Federal Republic of Germany. Data sheets often appear as regular features in each issue of a technical periodical, examples of titles including these features are *Air conditioning, heating and ventilating, EM & D: journal of engineering materials, components and design, Heating, piping and air conditioning, Machine design, Metal progress, Power, Tool and manufacturing engineer* and *Welding engineer.*

REPORT LITERATURE

A technical report, which may be produced within the research department of either an industrial firm, a trade or professional association, a government department or a university, is an account of work carried out on a research project by an engineer or group of engineers to convey information to technical management and other engineers working in the same field. Reports may be ' classified ' or ' unclassified ', classification in this context referring to the degree of secrecy of the information contained in the report which prevents its disclosure to unauthorised persons. The circulation of classified reports is severely limited, often to selected personnel within the producing organisation, while unclassified reports are usually generally available. Technical reports are also used to record information for patenting purposes, and to satisfy future information needs within the organisation producing the report. In large companies with extensive research programmes, the company's collection of its own technical reports will be one of the most important sources of technical information. This collection will represent the company's corporate store of the ' know-how ' developed by its technical personnel. The report can be either a final report, produced on completion of the project, or an interim report, produced to disseminate information about the progress of an investigation.

Reports represent an evolution from older methods of disseminating information: the writing up of laboratory notebooks, personal communication between scientists and publication in the primary journals. Technical reports are a relatively new form of publication and the growth of this literature has been parallel to the development of team research in the big science-based industries. Personal communication was an adequate method of disseminating information in a more leisurely age when scientists and engineers were able to know exactly who was working within their fields. Such conditions no longer obtain and research is now undertaken by teams working in literally thousands of laboratories throughout the world. In certain situations the report form has advantages over the periodical as a medium of communicating technical information. Reports are usually produced within the organisation sponsoring the research, by office duplicating processes, and are therefore well suited to the rapid dissemination of information, thus having a definite advantage over journal publication where delays of many months are often encountered in the publication of papers. The circulation of a report which contains classified research information can also be conveniently limited to those who have a need to know of the information disclosed, whereas with publication in the open literature unauthorised dissemination would be impossible to control. Much of the information which initially appears in the form of report literature is ultimately published in periodical or monograph form. This situation renders many technical reports obsolete within a period of three to four years, and therefore any substantial collection of report literature should be periodically ' weeded ' to avoid the duplication of material available in other forms. A measure of the development of the report as a form of technical literature is the recent estimate in the *Weinberg report* of the publication of 100,000 US government sponsored technical reports per annum.[6] The aeronautical and atomic energy industries have been developed almost exclusively on the information contained in report literature.

Technical reports are not usually listed in the book trade or national bibliographies, and many abstracting services give only superficial coverage of this form of literature. The identification of reports on a particular subject has, however, been facilitated in recent years by the publication of several specialised report abstracting services and bibliographies. In the United States *US Government research and development reports (USGRDR)* is a twice-monthly abstracting journal issued by the Clearinghouse for Federal Scientific and Technical Information (CHFSTI) (US Department of Commerce, National Bureau of Standards, Springfield, Virginia 22151), to announce the availability of government sponsored research and development reports released by the Department of Defense, the National Aeronautics and Space Administration and other government agencies. Arrangement is within twenty two categories (including aeronautics; materials; mechanical, industrial, civil and marine engineering; nuclear science and technology, etc) each of which is again subdivided, and each issue contains a report number index. All reports announced in *USGRDR* are indexed in *US Government research and development reports index* (*USGRDR*), a twice-monthly journal consisting of alphabetical subject, personal and corporate author, contract number and report number indexes with quarterly and annual cumulations. Copies of all reports listed in *USGRDR* are available from CHFSTI in the United States, and from the National Lending Library for Science and Technology in the United Kingdom. CHFSTI has also made available over 200 *Selective bibliographies* and *Catalogs of technical reports* covering many areas of engineering interest. A list of these publications is freely available from CHFSTI. The *Clearinghouse announcements in science and technology* service is a twice-monthly abstracting news service designed for rapid review of current reports in any of forty six fields, including aerodynamics and fluid mechanics, fuels and lubricants, materials, mechanical engineering, metals and alloys, etc. *Scientific and technical aerospace reports* (STAR) is also a twice-monthly abstracting service issued by the Scientific and Technical Information Division of the National Aeronautics and Space Administration covering world-wide report literature on space technology and aeronautics, and also NASA patents and patent applications. Arrangement is under thirty four subject categories, and each issue contains full subject, personal and corporate, and report number indexes. Retrospective searching is facilitated by the provision of quarterly and annual cumulated indexes. *Nuclear science abstracts* (*NSA*) is another twice-monthly abstracting service issued by the Division of Technical Information of the United States Atomic Energy Commission to announce the availability of unclassified reports published by the commission. *NSA* also covers world-wide journal and patent literature, books and monographs and the unclassified nuclear science and technology reports issued by other national atomic energy agencies. Arrangement is under broad subject groups, each of which is further sub-divided, and again subject, personal and corporate author and report number indexes which cumulate semi-annually and annually are provided.

British research and development reports (BRDR) is a monthly classified listing issued by the National Lending Library for Science and Technology (NLL) to promote the wider use of British reports. NLL seeks to include in BRDR any unclassified report covering any aspect of science, technology or the social sciences issued by any organisation within the United Kingdom. A photocopy of any report listed in *BRDR* is available on loan from NLL, and in addition this publication includes the names and addresses of British organisations issuing regular series of reports who are willing to make retention copies available to enquirers. NLL also maintains a loan collection of reports acquired from overseas agencies, including all reports issued by CHFSTI covering work carried out under US Government contracts, the reports issued by the national atomic energy authorities of over forty countries and a broad range of reports issued by foreign universities and industrial organisations. The Ministry of Technology Technical Reports Centre (TRC) at St Mary Cray, Orpington, Kent aims to collect all technical reports resulting from government sponsored research in the United Kingdom, and also those technical reports published overseas which are likely to be of interest to British industry, and then to make these reports available as widely as possible. *R & D abstracts*, which is published twice a month by TRC, includes abstracts of British and foreign reports within twenty two subject groups. Each issue contains an average of 300 abstracts and any abstract included may be obtained from TRC. The centre also operates an inquiry service which includes the compilation of bibliographies to customers' requirements. Unclassified reports issued by the United Kingdom Atomic Energy Authority are listed in the monthly *UKAEA list of publications available to the public.*

CONFERENCE PROCEEDINGS

The organisation of conferences at which papers are delivered and then subsequently published is a regular activity of most professional and research associations. Such papers often appear in the official journals of the sponsoring body, but in other cases they may be published in a collected form under a title such as *Proceedings of the nth conference on . . .* or *Transactions of the nth international meeting on . . . .* Although not all conference papers receive or merit publication, it should be noted that much useful information is lost because of the non-publication of the papers presented at some conferences.[7] It has also been demonstrated by ASLIB that the full influence of a conference is sometimes dissipated because of the difficulty of learning about its published proceedings through conventional bibliographies and abstracting services.[8] Several specialised publications have, however, been issued in the last few years in an effort to give information about forthcoming conferences and meetings, and also to record details of their published proceedings. These publications are of two kinds: calendars of meetings and bibliographies of publications.

The most comprehensive list of forthcoming international meetings is the Library of Congress *World list of future international meetings. Part I: science, technology, agriculture, medicine*. This is a monthly publication of which the March, June, September and December issues list all international meetings scheduled for the ensuing three years. The intervening issues list only new meetings and changes in those previously listed. This calendar is arranged by date of conference with indexes of subjects, sponsors and geographical locations. *Scientific meetings* is issued each quarter by the Special Libraries Association, and is an alphabetical and chronological listing of future regional, national and international scientific, technical and management conventions and symposia. Two new quarterly publications published by TMIS (79 Drumlin Road, Newton Centre, Massachusetts 02159) listing conferences and meetings in scientific, engineering and medical fields are *World meetings . . . United States and Canada* and *World meetings . . outside the United States and Canada*. Both include information on the content and locations of the meetings and details of proposed publications. A useful list of particular relevance to mechanical engineers is published in the United Kingdom by the Iron and Steel Institute. The *World calendar of forthcoming meetings: metallurgical and related fields* is a cumulative bulletin issued six times per year, and arranged chronologically by date of meetings with indexes of subjects, sponsoring bodies and locations. A significant feature of this calendar is the information it includes on forthcoming published proceedings, indicating the language, form and date of publication, etc. Other lists of national and international meetings are issued in the United Kingdom by the Ministry of Defence, Naval Scientific and Technical Information Centre and the International Scientific Relations Division of the Department of Education and Science.

Published conference proceedings can now be readily identified by consulting a number of current bibliographies. The Interdok Corporation (PO Box 81, Gedney Station, White Plains, NY 10605) issues *Interdok: directory of published proceedings, series SEMT: science, engineering, medicine, technology.* This monthly publication includes brief bibliographical details of conference proceedings published after January 1 1964, in chronological order by date of the original conference with indexes of subjects, sponsors, locations and editors. An annual cumulation of *Interdok* is available. *Current index to conference papers in engineering* is a monthly publication issued by CCM information Corporation (909 3rd Avenue, New York, NY 10022) which includes a conference data section giving information on the availability of preprints and proceedings and a subject and author index. In the United Kingdom the National Lending Library for Science and Technology issues the quarterly *Index of conference proceedings received by the NLL,* which covers all fields of the natural sciences, the technologies and the social sciences, and lists proceedings giving their title and date of conference only with a subject keyword index.

THESES

Theses are produced by research students to document a particular investigation in fulfilment of the requirements of a university or similar academic institution for a higher degree. These documents are usually made available to enquirers by the academic institutions, and are therefore a possible source of information on engineering subjects. In the United Kingdom ASLIB has published the annual *Index to theses accepted for higher degrees in the universities of Great Britain and Ireland* since 1950. This publication, which is somewhat slow to appear (the volume covering 1966-67 appeared in 1969), is a classified grouping of theses which incorporates sections on, for example, production engineering, machine tools and metallurgy. Author and subject indexes are provided in each annual volume and information on the availability of theses from each university is given. The American *Dissertation abstracts* changed its title to *Dissertation abstracts international* in 1969 to reflect its projected enlargement to include European theses. Section B *Sciences and engineering* is published monthly, and is restricted to material submitted by American, Canadian and, ultimately, European universities for publication by the publishers University Microfilms (300 North Zeeb Road, Ann Arbor, Michigan 48106). Within a classified arrangement each dissertation is represented by a detailed informative abstract and each issue is equipped with author and keyword indexes. Any title listed in *Dissertation abstracts international* is available from University Microfilms either on microfilm or as hard copy.

A complete inventory of all theses accepted by universities in the United States and Canada is published by University Microfilms in the annual *American doctoral dissertations*. The same organisation has also recently introduced the DATRIX service, a computer-based searching system which retrieves for clients details of the American theses published on any subject since 1938. Information on the theses published in overseas countries is often given in the country's national bibliography as in the case of *Bibliographie de la France* which issues a separate supplement covering dissertations. An important retrospective bibliography of theses is Marckworth, C M: *Dissertations in physics: an indexed bibliography of all doctoral dissertations accepted by American universities, 1861-1959*. Stanford University Press, 1961. Arrangement is by author, with an extensive keyword index, and the bibliography contains 8,400 titles. The preface states ' as different universities reflect the interests of their faculties and communities, some theses cited here are oriented primarily toward . . . electrical and mechanical engineering. In general, however, the major interest is that of the physicist.'

TRADE LITERATURE

Trade literature, which is issued by manufacturers to promote the sale and use of their products, will usually contain descriptive information on products and processes: details of their size and specifications, information on their installation, maintenance, use and applications. The form of this literature may vary from a single sheet covering a single component to the comprehensive, expensively produced catalogue covering a firm's entire range of products. Design and production engineers who need to know which components or instruments are available for a particular application will frequently need to refer to a collection of trade literature. This type of information is not available from other forms of literature. Trade literature can be acquired by scanning technical journals for manufacturers' advertisements and then requesting samples of literature by returning the pre-paid postcards which usually accompany the advertisements. Controlled-circulation journals are a particularly fruitful source for acquiring trade literature. Another approach is to compile a list of firms manufacturing particular articles of interest by scanning the classified sections of trade directories and then sending out a circular letter to these firms asking that the requesting organisation's name be placed on trade literature mailing lists. All trade literature should be date-stamped on receipt, as the date of publication is often omitted by the publisher and out-of-date product information is invariably worthless and misleading. Any collection of trade literature should also be regularly weeded to discard dated catalogues and to effect their replacement. The selection and weeding processes can be extremely time-consuming; there are however, in the United Kingdom and the United States a number of commercially based product information services which will provide up-to-date trade literature, usually prepackaged in microform on annual subscription. Technical Indexes (Index House, Ascot, Berkshire) offer services covering general engineering, production engineering, electronic engineering and chemical engineering. In each instance a periodically up-dated *Product data book* acts as a quick reference index to a 16mm roll microfilm file of product data compiled from the catalogues and literature issued by companies manufacturing products and components for consumption by British industry. Each microfile is revised every six months. Technical Indexes' services have recently been expanded to include supplementary microfiles of British standards and American military specifications and standards. TIM (Technical Information on Microfilm) (Wellington House, 125/130 Strand, London WC2) presents manufacturers' literature covering the fields of electronics, electrical and mechanical engineering components, materials and instruments on 6in × 4in microfiche with a capacity of 3,200 pages per fiche, this being made possible by the PCMI (photo-chromic micro-image) process which can produce reduction ratios of 1:150 and upwards. The TIM service also includes British and foreign patent specifications and standards, and all sections are revised every six months. In the United States *Thomas' register of American manufacturers* makes manufacturers' catalogues available on microfilm through its *Thomas Micro*

Catalogs Service (Thomas Publishing Co, 461 8th Avenue, New York, NY 10001). Sweet's Catalog Service's *Machine tool catalog file* (McGraw-Hill Co and F W Dodge Corp, 119 West 40th Street, New York, NY 10018) is a classified annual collection of manufacturers' catalogues issued as a bound volume.

GOVERNMENT PUBLICATIONS

Her Majesty's Stationery Office (HMSO) is the largest publisher in the United Kingdom, currently publishing about 6,000 items per year for parliament, government departments and public institutions. These publications can be divided into i) parliamentary publications, which are presented to parliament to cater for its information needs (House of Commons papers and bills, command papers, etc), or are issued to record its activity (parliamentary debates, statutes, etc); ii) non-parliamentary publications issued by HMSO on behalf of government departments, public institutions, nationalised industries, etc. Non-parliamentary publications vary in form enormously, from highly specialised technical reports issued by government research stations, to standard texts such as the Ministry of Defence Navy Department's *Applied mechanics,* and to popular manuals such as the Metropolitan Police Department's *Roadcraft.* In the case of non-parliamentary publications, the originating department is responsible for the editorial preparation of the document while HMSO determines its format and typography, decides on the number of copies to be printed and arranges for its distribution and sale.

Only a small proportion of HMSO publications appear in the *British national bibliography* and the book trade bibliographies, and thus to trace effectively the government publications which have been placed on sale the searcher must use the official indexes published by HMSO. The mimeographed *Daily list* covers HMSO publications as they appear, and this cumulates into the *Monthly catalogue,* each issue of which includes a descriptive loose-leafed insert giving short accounts of the most significant of each month's publications. The annual *Catalogue of government publications* cumulates the *Monthly catalogue* and is issued in the early months of the year following that which it covers. This *Catalogue* is paginated in consecutive sequences of five years to facilitate reference through the consolidated index to government publications which is issued quinquennially. The daily, monthly and annual lists are divided into parliamentary and non-parliamentary sections, the non-parliamentary publications being arranged by originating department. Subject, department and chairmen of committee indexes are included in the monthly and annual lists. The publications of the international bodies, such as OECD, NATO and EEC, for which HMSO acts as agent are included in both the daily and monthly lists, and *International organisations and overseas agencies' publications* is issued as a supplement to the *Annual catalogue*.

The publications of the various British government departments are documented in a series of over fifty *Sectional lists,* each of which lists all a department's current publications with a selection of important parliamentary publications covering its subject. These lists are revised periodically and are available free of charge from HMSO. Among the lists of interest to the mechanical engineer are those for the Ministry of Technology (3) which includes technical reports and notes published for the ministry's research stations, the Aeronautical Research Council (8) which includes the *Reports and memoranda* and the *Current papers* reports series, and the Atomic Energy (63) list which includes a classified listing of reports issued by the United Kingdom Atomic Energy Authority.

The *Monthly catalog of United States government publications* is issued by the Superintendent of Documents (United States Government Printing Office (USGPO), Washington, DC 20402), and embraces all government publications, including congressional and departmental and bureaux publications. Departmental publications are listed under the name of the originating departments, which are arranged alphabetically according to the significant word in the department's name. An annual index is issued with each completed volume. The USGPO *Price lists,* which are similar to the British *Sectional lists,* are individual subject catalogues of current United States government publications covering specific subject fields. Those currently available which are of direct interest to the mechanical engineer include *Transportation* (25), *Scientific tests, standards* (64), *Aviation* (79), *Atomic energy* (84), and *Consumer information* (86).

REFERENCES

1 Newby, F: *How to find out about patents*. Pergamon Press, 1967. pp 111-116.

2 Houghton, B: *Technical information sources*. Bingley, 1967. p 39.

3 *Ibid* pp 45-46.

4 *Standardization*, (41) January 1960, p 10.

5 Houghton, B: *Op cit* pp 70-72.

6 *Science government and information*. United States Government Printing Office, 1963. pp 39-40.

7 Liebesney, F: *Proceedings of the international conference on scientific information, Washington, DC Nov 16-21, 1958*. National Academy of Sciences, National Research Council, 1959. Vol I, pp 474-479.

8 'Availability of scientific conference papers and proceedings.' *Unesco bulletin for libraries, 16*(4) Jul/Aug 1962, pp 165-176.

CHAPTER 12

REFERENCE TOOLS
DICTIONARIES, ENCYCLOPEDIAS, HANDBOOKS, ETC

The scope of this chapter includes general reference works on science and engineering, in addition to those more specific reference tools covering mechanical engineering and its component and related subjects. The phrase 'reference work' is taken here to include encyclopedias and dictionaries, handbooks and data books, volumes of tables and formulae, and, finally, directories of the various industries.

The contents of an encyclopedia are usually arranged in an alphabetical sequence under the names of things and concepts, with each entry giving a concise account of its subject. Encyclopedias are essentially sources of preliminary knowledge to which the enquirer can turn for an introductory explanation of a principle with which he is unfamiliar. The encyclopedia is usually compiled under the direction of an editor or editorial board, by a team of specialist contributors, each responsible for the entries in a particular sector of the subject field being covered. Some encyclopedias are of additional value in that they provide further references to selected works on the subjects which they cover. A good example of an encyclopedia providing references to further reading is that covering technology in general, the *McGraw-Hill encyclopedia of science and technology*.

Technical dictionaries provide only brief definitions of the terms in current use in the various disciplines and industries, although some, *eg* Horner's *Dictionary of mechanical engineering terms*, are encyclopedic in nature, as they include extended treatment of important subjects.

Handbooks and data books can be primarily considered as the working reference tools of practising engineers. They consist of concise digests of information on the basic principles and methods used in the various branches of engineering. In addition to their summaries of basic concepts they invariably contain a wide range of mathematical tables and data of use to design engineers, practical reference informa-

7

193

tion for the plant and maintenance engineer, summaries of standards and specifications, and information on the properties and applications of materials. Their place is as much in the laboratory and workshop as in the library, and they are of particular value to engineers who do not have access to a large library. Their value is, however, not limited to such situations: they are in constant reference use even in the largest libraries, as they present in convenient, often tabular, form information which may have been previously published in, for instance, primary periodicals or technical reports, and which therefore may otherwise be buried and scattered throughout the literature. Such information would in many cases be difficult to retrieve if it were not for the existence of these compilations. There are often two or even three handbooks covering the same subject area, and the inclusion of more than one title in the following listings is because, although much of the information they include is common to each volume, one handbook contains tables or data not present in the other.

Included in the sections on directories and yearbooks are the publications giving alphabetical and classified listings of manufacturers in general, and of manufacturers in particular industries, and also the catalogues produced by trade and other organisations which give descriptive information on the specifications of manufacturers' products. Both these types of publication often include other valuable commercial and technical information, such as lists of trade names and information on the activities of and services available from trade and research organisations in their field.

The following publications are useful in tracing other reference works in subjects related to mechanical engineering:

Walford, A J (ed): *Guide to reference material. Vol I: science and technology*. London, Library Association, second edition 1966.

Winchell, Constance M (ed): *Guide to reference books*. Chicago, American Library Association, eighth edition 1967.

TABLES

Fletcher, A and others: *An index of mathematical tables*. Oxford, Blackwell for Scientific Computing Services Ltd, second edition, two volumes 1962. Vol I: index according to functions; vol II: bibliography.

Schutte, Karl (ed): *Index of mathematical tables from all branches of science.* Munich, Oldenbourg, second edition 1966.

Harvey, Anthony P (ed): *Directory of scientific directories: a world guide to scientific directories including medicine, agriculture, engineering, manufacturing and industrial directories.* Guernsey, Francis Hodgson Ltd, 1969. An international directory arranged alphabetically by country.

Henderson, G P and Anderson, I G, *Current British directories.* London, Jones & Evans Bookshop, sixth edition 1970-71. Pt I: local directories; pt II: specialised directories.

Klein, Bernard (ed): *Guide to American directories: a guide to the major business directories of the United States covering all industrial, professional and mercantile categories.* New York, McGraw-Hill, seventh edition 1968. Arranged alphabetically by subject.

Special Libraries Association: *Guide to special issues and indexes of periodicals.* Edited by Doris B Katz and others. New York, SLA, 1962. A useful guide arranged alphabetically by title of periodical listing those periodicals which regularly publish buyers' guides, directories, etc as special issues.

PHYSICS

Handbooks and data books

Condon, E V & Odishaw, H: *Handbook of physics.* New York, McGraw-Hill, second edition 1967. Attempts to select from the vast literature of physics of materials ' what every physicist should know '. Contents: mathematics, mechanics of particles and rigid bodies, mechanics of deformable bodies, electricity and magnetism, heat and thermodynamics, optics, atomic physics, solid state, nuclear physics.

Gray, D W (ed): *American Institute of Physics handbook.* New York, McGraw-Hill, second edition 1963. A working tool for those using physical methods in engineering and research comprising of physical data in tabular and graphical form. Covers: aids to computation, mechanics, acoustics, heat, electricity and magnetism, optics, atomic and molecular physics, nuclear physics, solid state physics.

A classified listing of reference sources follows:

Dictionaries and encyclopedias

Ballentyne, D W G and Walker, L E Q: *Dictionary of named effects and laws in chemistry, physics and mathematics*. London, Chapman and Hall, second edition 1961. An alphabetical listing of scientists whose names have become linked with scientific laws of effect, including a brief description of each law.

Chambers' technical dictionary. Edited by C F Tweney and L E C Hughes. London, Chambers, third edition revised with supplement 1958. The best handy one volume technical dictionary available. Aims to give definitions of the terms which are of importance in pure and applied science including all branches of engineering and construction and the larger manufacturing industries and skilled trades.

Crispin, F S: *Dictionary of technical terms*. Milwaukee, Bruce, tenth edition 1964. Definitions of commonly used expressions in aeronautics, architecture, woodworking and building trades, electrical and metalworking trades, printing, chemistry, plastics, etc.

International encyclopedia of science. Edited by J R Newman. London, Nelson, four volumes 1965. First published in the United States as the *Harper encyclopedia of science*. A work aimed at the non-specialist which consists mainly of longer integrated articles rather than of a multitude of shorter dictionary type entries. The work is adequately illustrated and bibliographies are contained in volume 4.

Jones, F D and Schubert, P B: *Engineering encyclopedia*. New York, Interscience, third edition 1963. A condensed encyclopedia and mechanical dictionary for engineers, mechanics, technical schools and industrial plants and public libraries giving the basic facts on 4,500 important engineering subjects.

McGraw-Hill encyclopedia of science and technology. New York, McGraw-Hill, fifteen volumes 1965. Attempts ' to provide the widest possible range of articles that will be intelligible to a person of modest technical training who wants to obtain information outside his particular field of specialisation.' Consists of 7,200 articles whose length varies from several pages to a few lines. Bibliographies are appended to many articles and the work contains numerous tables, diagrams and photographs. The *McGraw-Hill yearbook of science and technology* includes both feature articles on new developments and reviews of progress in the various fields of science and technology during the past year.

Scott, A A H (ed): *Engineering outlines.* London, Macmillan, 1967-. An annual cumulation of the weekly *Engineering outline* features taken from the periodical *Engineering*. These volumes are useful reference tool for engineers who wish to keep up-to-date with techniques outside their own specialisations. Each outline includes a description of basic principles, applications, advantages and limitations, sources of further information.

Van Nostrands' scientific encyclopedia. New York, Van Nostrand, fourth edition 1968. The standard one volume encyclopedia of science and engineering. Incorporates two-tier definitions: initially a simple definition expressed in the plainest terms followed by a more detailed explanation of the topic.

Handbooks and data books

Esbach, O V: *Handbook of engineering fundamentals.* New York, Wiley, second edition 1952. Attempts to bring together the most important basic facts and principles and the fundamental laws of science upon which technological and engineering development depends. Covers: mathematics, physics, chemistry, properties and uses of engineering materials, mechanics of solids and fluids, commonly used mathematical and physical tables.

Grazda, E E and others: *Handbook of applied mathematics.* Princeton NJ, Van Nostrand, fourth edition 1966. Broad reviews of basic mathematics, *eg* algebra, trigonometry, calculus, etc are followed by chapters on methods and calculations applicable to particular industries, *eg* sheet metal, machine shop practice, heating and ventilating, etc.

Handbook of chemistry and physics: a ready reference book of chemical and physical data. Cleveland, Ohio, Chemical Rubber Publishing Co, forty ninth edition 1969. Revised almost annually since 1913. Probably the most useful general purpose handbook covering chemistry, physics and related sciences.

Kemp's engineers year book. London, Morgan-Grampian, two volumes. The chief British compilation of data, formulae, tables and descriptions of basic principles covering every branch of engineering. An index of over 12,000 entries included in volume 2.

Molesworth's handbook of engineering formulae and data. Edited by

A P Thurston. London, Spon, thirty fourth edition 1951. A successor to *Molesworth's pocket book of engineering formulae and memoranda,* first published in 1862. Contains sections covering general units, conversion factors, materials, civil and general engineering, mechanical engineering, electrical engineering.

Perry, R H: *Engineering manual: a practical reference book of data and methods in architectural, chemical, civil, electrical, mechanical and nuclear engineering.* New York, McGraw-Hill, second edition 1967. A compilation of commonly used formulae, data and methods for non-detailed design in any engineering area. Section 2, the engineering core, attempts to make accessible information applicable to all fields of engineering, *eg* fluid flow, heat transfer, thermodynamics, etc.

Potter, J H: *Handbook of the engineering sciences. Vol I: basic sciences 1967. Vol II: applied sciences 1968* Princeton, NJ, Van Nostrand. Aims to present the fundamental considerations of the engineering sciences ' on a level approximating that of the first year graduate student in engineering '. Vol I, seven sections: mathematics, physics, chemistry, graphics, statics, theory of experiments, mechanics. Vol 2 aims to provide all engineers with comprehensive information on the specific engineering sciences, with eighteen sections, including heat and mass transfer, turbomachinery, materials science, machine elements, etc.

Souders, M: *The engineers companion: a concise handbook of engineering fundamentals.* New York, Wiley, 1966. Intended as a comprehensive, concise and portable reference book of tables and data which omits considerations of techniques and examples of applications. Contents covers: mathematics, mechanics, fluid mechanics, thermodynamics, heat transfer, electricity and magnetism, nuclear physics, engineering economics, mathematical and physical tables.

Symbols, units, formulae, etc

Arnell, A: *Standard graphical symbols: a comprehensive guide for use in industry, engineering and science.* New York, McGraw-Hill, 1963.

Gieck, K: *A collection of technical formulae.* English translation by F B Catty. London, Bailey, 1967. Attempts to provide engineers with a brief guide to the more important technical and mathematical formulae.

198

Hvistendahl, H S: *Engineering units and physical quantities.* London, Macmillan, 1964. Covers the various systems of units used by engineers.

Jerrard, H G and McNeil, D B: *A dictionary of scientific units including dimensionless numbers and scales.* London, Chapman and Hall, 1963. Definitions and historical details of some four hundred units normally employed by scientists.

Polon, D D: *Encyclopedia of engineering signs and symbols.* New York, Odyssey Press, 1965. A compilation of the graphic symbols, signs and descriptions in general use throughout the engineering sciences.

Tables

American society of mechanical engineers: *ASME handbook. Engineering tables.* New York, McGraw-Hill, 1956. Aims to collect into one volume those tables which are generally recognized as standard and which are often required by engineering designers but which are not commonly found in handbooks. The subjects of tables included are for example bearings, gearing, fasteners, springs, gaskets, seals, etc.

Grytz, E: *Technical tables.* Translated from the German by H Liebscher. Leipzig, VEB Edition, 1963. Tabular information and data covering mathematics; physics; mechanical engineering; electrical engineering; chemical engineering; materials.

Handbook of mathematical tables; supplement to handbook of chemistry and physics. Cleveland, Ohio, Chemical Rubber Publishing Co, 1962. Mathematical tables for computational work in physics, chemistry and engineering.

Kaye, G W C and Laby, T H: *Tables of physical and chemical constants and some mathematical functions.* London, Longmans, thirteenth edition 1966. Contents: general physics, chemistry, atomic physics, mathematical tables.

Landolt, H H and Bornstein, R: *Zahlenwerte und funktionen aus physik, chemie, astronomie, geophysik und technik.* Berlin, Springer Verlag, sixth edition, four volumes issued in twenty parts 1950-.

Landolt, H H and Bornstein, R. *Zahlenwerte und funktionen aus naturwissenschaften und technik. Neue serie.* Berlin, Springer Verlag, 1961-. The above works constitute the most comprehensive sets of tables covering physics and related disciplines yet published. The new series, which is being issued to keep the sixth edition up to date, is easier to use in that it incorporates added title pages and tables of contents, etc in English.

Naft, S and De Sola, R: *International conversion tables*. London, Cassell, second edition 1965. Contents: conversion factors, conversion tables, compound conversion, conversion measures in various industries, geography of weights and measures.

National Research Council: *International critical tables of numerical data, physics, chemistry and technology; prepared under the auspices of the International Research Council and the National Academy of Sciences by the National Research Council of the United States of America.* New York, McGraw-Hill, seven volumes and index 1926-1933. Critical tables in that the cooperating experts were requested in each instance to provide the ' best ' value which they could derive from all the information available, together where possible with an indication of its probable reliability. Full bibliographical references are given to the origin of the data.

Directories

Engineer buyers' guide. London, *The engineer,* annual. Contents: manufacturers' addresses, UK agents for foreign manufacturers, trade names, classified buyers' guide.

Kelly's manufacturers' and merchants' directory including industrial services. London, Kelly's Directories. *Vol 1: Great Britain, Northern Ireland, Republic of Ireland. Vol 2: Europe, Africa, America, Asia, Oceania.* Annual. Vol 1 includes an alphabetical list of firms in addition to classified listings; Vol 2 includes classified lists only.

Machinery's annual buyers' guide. London, Machinery Publishing Co. Trade names, classified list of manufacturers, addresses.

McRae's blue book. Western Springs, Illinois, McRae's Blue Book, four volumes, annual. Vol 1 alphabetical list of US firms, vols 2-4 classified list of US manufacturers.

Ryland's coal, iron, steel, tinplate, metal, engineering, foundry, hardware and allied trades directory, 1968-69. London, Fuel and Metallurgical Journals, two volumes 1968. Vol 1 alphabetical list of firms and geographical lists of firms in UK, vol 2 classified trades.

Thomas' register of American manufacturers. Seven volumes, annual.

Vols 1-6 companies classified by trade, vol 7 alphabetical list of _ American companies.

UK Kompass: register of British industry and commerce. London, Kompass Register Ltd, sixth edition, three volumes 1968. Vol 1 names and addresses of UK companies, vol 2 classified listing of companies by products and services, vol 3 detailed information on leading companies arranged alphabetically by city or town.

MECHANICAL ENGINEERING

Dictionaries and encyclopedias

Audel's new mechanical dictionary for technical trades containing 11,000 definitions of commonly used terms in mechanical trades, physics, chemistry, electricity, etc. New York, Audel, 1960. Generous definitions intended for the practical man.

Auger, C P: *Engineering eponyms: an annotated bibliography of some named elements, principles and machines in mechanical engineering.* London, Library Association, 1965. Gives an alphabetical list of eponyms with a brief description and a list of references to sources of further information for each.

Del Vecchio, A: *Dictionary of mechanical engineering.* New York, Philosophical Library, 1961. Brief definitions of terms used in mechanical engineering and related fields such as electricity, heat treatment of metals, welding and mathematics.

Horner, J G: *Dictionary of mechanical engineering terms.* Ninth edition revised and enlarged by G K Grahame-White. London, Technical Press, 1967. Detailed definitions used in the theory and practice of mechanical engineering. Pt 1 modern terms, pt 2 basic terminology.

Nayler, J L and Nayler, G H F: *Dictionary of mechanical engineering.* London, Newnes, 1967. An illustrated dictionary including concise definitions of terms covering the production of the means for and the utilisation of mechanical power in engines, transport and mechanics but ignoring many allied fields such as foundry practice, metallurgy, metrology and welding.

Universal encyclopedia of machines or how things work. London, Allen & Unwin, 1967. Simple explanations of both simple mechanical functions and more complex industrial processes.

Handbooks and data books

Baumeister, T and Marks, L S: *Standard handbook for mechanical engineers*. New York, McGraw-Hill, second edition 1967. Best one-volume reference tool for mechanical engineers. Includes basic mathematical tables and summaries of essential data and principles on every aspect of mechanical engineering. Contents: mathematical tables and weights and measures, mathematics, mechanics of solids and fluids, heat, strength of materials, materials of engineering, fuels and furnaces, machine elements, power generation, materials handling, transportation, building construction and equipment, shop processes, pumps and compressors, electrical and electronics engineering, instruments and controls, industrial engineering, refrigeration engineering. Detailed index of over 12,000 entries. Bibliographical references given at the beginning of each section.

Collins, A T: *Newnes' engineer's reference book*. London, Newnes, tenth edition 1965. Planned as an authoritative work of reference on all branches of mechanical engineering. Attempts to assemble in a single volume data, tables and information formerly scattered through several volumes. Essential day-to-day information is contained in *Newnes' engineer's manual for designers, draughtsmen, toolmakers, turners, fitters, mechanics, erectors and students*. 1964.

Flugge, W: *Handbook of engineering mechanics*. New York, McGraw-Hill, 1962. Seven sections covering: mathematics, mechanics of rigid bodies, theory of structures, elasticity, plasticity and viscoelasticity, vibrations, fluid mechanics.

Fowler, W H: *Fowler's mechanical engineers' pocket book*. Manchester, Scientific Publishing Co. A concise data book issued annually which includes basic tables and brief information on: boilers, fuels, steam engines, valves and valve gear, gas engines, oil engines, hydraulics, gearing, bearings, lubrication.

Fowler, W H: *Fowler's mechanics' and machinists' pocket book: a synopsis of practical rules for fitters, turners, millwrights, erectors, pattern makers, foundrymen, draughtsmen, apprentices, students, etc.* Manchester, Scientific Publishing Co. annual.

Greenwood, D C: *Mechanical details for product design*. New York, McGraw-Hill, 1964. A selection of illustrated mechanical design aids taken from the pages of *Product engineering*, which attempts to provide 'ready-made experience in product design' by summarising the state of the arts. Scope includes basic general design, control and materials handling, fastening and joining, hydraulics and pneumatics, mechanical movements and linkage, mechanical power transmission, spring devices, welding and brazing.

Oberg, E and Jones, F D: *Machinery's handbook: a reference book for the mechanical engineer, draftsman, toolmaker and machinist*. New York, Industrial Press, eighteenth edition 1968. Data collected and presented in a form to meet the requirements of the design and production departments of both large and small manufacturing plants, the jobbing trade and the technical school. Information given on both American and British standards.

'Power', editorial staff: *Power generation systems*. New York, McGraw-Hill, 1966. A compilation of special reports previously published in *Power*, including energy systems, steam generation, steam turbines, gas turbines, diesel engines, etc.

Tables

Camm, F J: *Workshop calculations: tables and formulae for draughtsmen, engineers, fitters, turners, mechanics, patternmakers, erectors, foundrymen, millwrights, and technical students*. London, Newnes, fourteenth edition 1963.

Elzanowski, Z: *Mechanical engineers reference tables*. London, Iliffe, 1966.

Directories

American Society of Mechanical Engineers: *Mechanical engineers' catalog and plant directory*. New York, the Society. An annual listing of over 3,800 manufacturers and products giving technical information and specification data on the products.

British Mechanical Engineering Federation: *Mechanical engineering directory and buyers' guide*. London, the Federation, biennial.

Mechanical world year book. Manchester, Emmott and Co. An annual which, in addition to a classified buyers' directory of the British mechanical engineering industry, includes basic tables and information on materials, processes, machine tools, etc.

Dictionaries

Adams, F B: *Aeronautical dictionary*. Washington, United States Government Printing Office, 1959. An illustrated dictionary with detailed explanatory definitions.

McLaughlin, C: *Space age dictionary*. Princeton, NJ, Van Nostrand, second edition 1963. Simple non-technical definitions of important concepts in the fields of rockets, missiles, launch-vehicles, satellites and space flight.

Moser, R C: *Space age acronyms: abbreviations and designations*. New York, Plenum Press, second edition 1969. Includes official US Air Force and Army abbreviations and acronyms, in addition to foreign acronyms of military and commercial aircraft manufacturers.

Nayler, J L: *Dictionary of aeronautical engineering*. London, Newnes, 1959. An illustrated dictionary of concise definitions covering aeronautical engineering, and also related subjects such as artificial satellites, guided missiles and rockets.

Nayler, J L: *Dictionary of astronautics*. London, Newnes, 1964.

Handbooks and data books

Koelle, H H and Braun, W von: *Handbook of astronautical engineering*. New York, McGraw-Hill, 1961. A summary of technical and engineering data organised into six sections covering: fundamentals of astronautical engineering, astrodynamics, astrionics, propulsion systems, space vehicles, space flight operations.

Morrison, R B and Ingle, M J: *Design data for aeronautics and astronautics: a compilation of existing data*. New York, Wiley, 1962. Basic information needed for scientific and engineering work in aeronautics and astronautics. Covers: atmospheric properties, thermodynamic properties of gases, fluid flow properties, aerodynamics, aerothermodynamics, performance, materials, aerothermochemistry, human tolerances.

Overbey, C A: *Aircraft and missile design and maintenance handbook*. London, Macmillan, 1960. A practical handbook for mechanics, inspectors, draughtsmen, designers, modification engineers, etc working in aircraft factories and airports covering standard methods of equipment installation and maintenance.

Wukelie, G E: *Handbook of Soviet space-science research*. New

York, Gordon and Breach, 1968. Covers Soviet space research from Sputnik I to December 1966; rockets, spacecraft and results are described and tabulated.

Directories

'Aeroplane' directory of British aviation including who's who in British aviation. London, Temple Press, 1949-. An annual directory of British aviation covering associations, manufacturers, the aeronautical press, commercial aviation, aeronautical training, etc.

Aerospace yearbook. Official publication of the Aerospace Industries Association of America Inc. Washington, American Aviation Publishers Inc, 1919-.

Aerospace Industries Association of America Inc: *Aerospace facts and figures,* Fallbrook, California, Aero Publishers. An annual publication giving statistical and general information on aircraft production, missile and space programmes, research and development, etc.

Aircraft and Missile Magazine: *Missile handbook: United States missile projects, missile directory and data, space programmes and missiles directory of contractors.* New York, Association of Missile and Rocket Industries, 1960. Kept up to date by supplements.

Jane's all the world's aircraft. London, Sampson Low, 1909-. An illustrated annual record of aviation development and progress. Arranged alphabetically by country, gives details of civil and military aircraft, airships, drones, sailplanes, military missiles, research rockets and space vehicles and aero-engines.

Wilkinson, P H: *Aircraft engines of the world.* Washington, Paul H Wilkinson, 1941-. An annual volume arranged in sections covering gas turbines and rocket jets, reciprocating engines, equipment and materials. 1966/67 edition contains data on some 177 gas turbines and seventy three reciprocating engines from fifteen countries.

World aviation directory; listing aviation companies and officials: covering the United States, Canada and 123 countries in Europe, etc. Washington, American Aviation Publications Inc.

ALUMINIUM

Aluminum Company of America: *Aluminum.* Edited by K R Van Horn. Novelty, Ohio, American Society for Metals, three volumes 1966. Vol 1: properties, physical metallurgy and phase diagrams; vol 2: design and applications; vol 3: fabrication and finishing.

Dictionaries and encyclopedias

Kutateladze, S S and Borishanskii, V M: *Concise encyclopaedia of heat transfer*. Oxford, Pergamon Press, 1966. A summary of existing knowledge and a collection of data and formulae used in calculations on all types of heat transfer problems.

Handbooks and data books

Boelter, L M K and others: *Heat transfer notes*. New York, McGraw-Hill, 1965. Covers: fundamental concepts, classical methods and detailed developments in mathematical analysis.

Fraas, A P and Ozisik, M R: *Heat exchanger design*. New York, Wiley, 1965. Basic theory, practical design information and analytical techniques necessary to cope with the engineering problems connected with a wide range of heat transfer equipment.

Geiringer, P L: *Handbook of heat transfer media*. New York, Reinhold, 1962. Produced for the mechanical and chemical engineer engaged on high temperature heating problems. Includes engineering data on heat transfer fluids used throughout the world.

Tables

Haywood, R W: *Thermodynamic tables in Si (metric) units, with conversion factors to other metric and British units*. Cambridge University Press, 1968. Thermochemical tables, steam tables, refrigerant tables, air at low temperature, transport properties of various fluids.

Mayhew, Y R and Rogers, G F C: *Thermodynamic properties of fluids and other data*. Oxford, Blackwell, 1962.

See also ENGINE DESIGN AND APPLICATION; HEATING AND VENTILATING ENGINEERING; REFRIGERATION ENGINEERING; STEAM ENGINEERING

AUTOMATION AND INSTRUMENTATION

Handbooks and data books

Carroll, G C: *Industrial instrument servicing handbook*. New York, McGraw-Hill, 1960. Servicing and maintenance information on modern industrial process measuring and control instruments manufactured by forty one companies.

Considine, D M: *Process instruments and control handbook*. New York, McGraw-Hill, 1957. A guide to the selection of instruments and automatic control covering operating and design fundamentals of equipment used in the process fields.

Considine, D M and Ross, S D: *Handbook of applied instrumentation.* New York, McGraw-Hill, 1964. Companion volume to *Process instruments and control handbook* giving data on instruments and their applications in all major fields.

Grabbe, E M and others: *Handbook of automation, computation and controls.* New York, Wiley, three volumes 1958-1961. Attempts to provide practical design data for research, development and design in feedback control, computers, data processing, control components and control systems. Vol 1 control fundamentals; vol 2 components and data processing; vol 3 systems and components.

Kallen, H P: *Handbook of instrumentation and controls: a practical manual for the mechanical services covering steam plants, power plants, heating systems, air conditioning systems, ventilation systems, diesel plants, refrigeration, water treatment.* New York, McGraw-Hill, 1961.

Leslie, W H P: *Numerical control users' handbook.* Maidenhead, McGraw-Hill, 1970.

Truxal, J G: *Control engineers' handbook: servomechanisms, regulators and automatic feedback control systems.* New York, McGraw-Hill, 1958. Basic information on components and design technique in feedback control systems. Emphasis on components, including electromechanical, mechanical, hydraulic, pneumatic, electronic and magnetic components. Covers working principles, applications and limitations of components.

Directories

British instruments directory and data handbook, 1965. London, Scientific Instrument Manufacturers Association, etc, 1965.

Control applications guide. New York, Automatic Control Magazine. An annual listing of American manufacturers and products in the fields of instrumentation and automatic control.

Control industry guide and digest. London, Morgan Brothers, 1962.

Institute of Physics and the Physical Society: *The physics exhibition.* London, the Institute. An annual descriptive catalogue of significant new British scientific instruments.

Instrument manual. Third edition edited by J T Miller. London, United Trade Press, 1960. Sixteen chapters, each surveying British instruments in a specific field, *eg* engineering precision instruments and gauges, aeronautical instruments, etc, with each including a bibliography and buyers' guide.

See also METROLOGY

AUTOMOBILE ENGINEERING
Handbooks and data books

Autocar handbook. London, Iliffe, twenty third edition 1969. A popular handbook covering the fundamental principles on which car design is based.

Coker, A J and Lancaster, G H: *Automobile engineers' reference book: a comprehensive work of reference providing a summary of the latest practice in all branches of automobile engineering.* London, Newnes, third edition 1959.

Murphy, P A and others: *Chilton's auto repair manual, 1968.* Philadelphia, Chilton, 1967. A compact guide to the servicing of American cars from 1960 to 1968, covering all systems of each model including detailed instructions for removal and installation.

Newton, K and Steeds, W: *The motor vehicle.* London, Iliffe, eighth edition 1966.

Servicing guide to British motor vehicles. London, Trader Publishing Co, eight volumes 1961-1965. Detailed servicing instructions for British motor cars and commercial vehicles.

Society of Automotive Engineers: *SAE handbook.* The Society, 1926-. An annual collection of SAE standards, recommended practices and information reports covering: ferrous metals, non-ferrous metals, non-metallic materials, threads, fasteners and common parts, electrical equipment and lighting, power plant components and accessories, passenger cars, trucks and buses, marine equipment.

Society of Motor Manufacturers and Traders: *Standards and technical information for the British automobile industry.* London, the Society, two volumes 1968. Standards, data sheets and information sheets for use by the automobile industry.

Directories
Automotive industries products guide. Philadelphia, Chilton Co, annual.

National automotive directory. Atlanta, WRC Publishing Co, annual.

Society of Motor Manufacturers and Traders: *Buyers' guide to the motor industry of Great Britain.* London, the Society, 1969. An annual including classified listings of firms, address list and trade names.

Trader handbook: a legal, technical and buying guide for the motor, motor cycle and cycle trades. London, Iliffe. An annual including car specifications and servicing data in addition to a buyers guide and address list.

BEARINGS
'*Machine design.' The bearings book.* Cleveland, Ohio, Penton Publishing Co, 1961. Covers common bearing design and selection problems and includes a manufacturers and products directory.

CASTING see FOUNDRY PRACTICE

COMPRESSED AIR see FLUID POWER AND PNEUMATICS

COPPER
Handbooks and data books
British Bronze and Brass Ingot Manufacturers' Association: *Bronze and brass ingot manual.* Birmingham, the Association, third edition 1966. Covers: general metallurgical information, specifications, castings, inspection, recommended uses, equivalent specifications.

Copper Development Association: *Copper and copper alloy data.* London, the Association, 1968. Physical, mechanical and other properties of the common copper-base materials.

Copper Development Association: *Standard handbook for copper and copper alloy wrought mill products.* New York, the Association, fifth edition 1964. Standards for wrought mill products of copper and copper alloys including plate, sheet, strip, rod, bar, wire, pipe, tube and shapes.

Yorkshire Engineering Supplies Ltd: *Bronze: a reference book.* Leeds, 1962.

CORROSION

Handbooks and data books

Anti-corrosion manual. London, Scientific Surveys Ltd, fourth edition 1962. Sections covering: corrosion in industry, resistant materials and their applications, preparatory treatment for protective coatings, protective coatings, mastics, corrosion testing, cathodic protection, etc.

Burns, R M and Bradley, W W: *Protective coatings for metals.* New York, Reinhold, third edition 1967. Survey of the latest information on metallic and organic protective coatings and their composition, structure, applications and evaluation.

Rabald, E: *Corrosion guide.* New York, American Elsevier, second edition 1968. A list of corrosive agents with data on their effect on materials.

Uhlig, H H: *Corrosion handbook.* London, Chapman and Hall, 1948.

DIESEL ENGINES see ENGINE DESIGN AND APPLICATION

ELECTROPLATING see METAL FINISHING

ENGINE DESIGN AND APPLICATION

Handbooks and data books

Audel's diesel engine manual. New York, Audel, 1962. A concise treatment of the theory, operation and maintenance of modern diesel engines.

Church, A H and Gartmann, H: *De Laval handbook; an engineering data book for users of pumps, turbines, compressors and gears.* Trenton, NJ, De Laval Steam Turbine Co., second edition 1955.

Judge, A W: *High speed diesel engines: with special reference to automotive, stationary and marine types.* London, Chapman and Hall, sixth edition 1967.

Sawyer, J W: *Gas turbine engineering handbook.* Stamford, Connecticut, Gas Turbine Publications, 1966. Technical data covering design, manufacture, testing, selection, operation and maintenance of gas turbines.

Schmidt, F A F: *The internal combustion engine.* Translated by

R S W Mitchell and J Horns. London, Chapman and Hall, third edition 1965. Pt 1 the reciprocating engine; pt 2 gas turbines.

Simmons, C R: *Gas turbine manual*. London, Temple Press, third edition 1968. An international review of current gas turbine practice in industrial, marine and transport applications. One section covers details of every known British gas turbine, and included as an appendix is a world directory of gas turbine manufacturers.

Southerton, R C: *Oil engine manual*. London, Temple Press, seventh edition 1964. Design and operation of oil engines for industrial use, railway and road transport.

Thomson, W R: *Preliminary design of gas turbines*. London, Emmott, 1963. Covers methods for designing the components of a gas turbine plant—compressor, turbine, heat exchanger, ducting.

Directories

British Internal Combustion Engine Manufacturers' Association : *British diesel engine catalogue*. London, the Association, sixth edition 1965. Specifications of and data on oil engines of compression ignition and associated types, also gas turbines for industrial, railway traction and marine duties made by members and other firms.

See also AERONAUTICAL AND AEROSPACE ENGINEERING; APPLIED HEAT; AUTOMOBILE ENGINEERING; LOCOMOTIVE ENGINEERING; MACHINE DESIGN; MARINE ENGINEERING; POWER TRANSMISSION.

FABRICATION AND JOINING

Handbooks and data books

American Welding Society: *Brazing manual*. New York, the Society, second edition 1963.

American Welding Society: *Soldering manual*. New York, the Society, 1959.

Bakish, R and White, S S: *Handbook of electron beam welding*. New York, Wiley, 1964. Covers the theory of electron beam generation; details of various units generally available; the physical and mechanical metallurgy of joints; tooling and estimation costs.

Clements, R: *Manual of light production engineering. Vol 1 : assembly methods and types of fasteners*. London, Business Books Ltd, 1968. Survey of the latest methods of assembly and fastening now in use in industry; covers: feeding devices, clamping devices, adhesives, welding, brazing, soldering, screws, nuts, bolts, mechanical joints.

Del Vecchio, E J: *Resistance welding manual.* Philadelphia, Resistance Welder Manufacturers' Association, two volumes 1956. Section 1 processes; 2 materials. Appendices of definitions, symbols, tables and bibliography.

Laughner, V H and Hargan, A D: *Handbook of fastening and joining of metal parts.* New York, McGraw-Hill, 1956. Aims to provide the design engineer with data on all known methods of joining metal parts. Covers standards, description of types, advantages, limitations and applications of methods.

Lincoln Electric Co: *Procedure handbook of arc welding design and practice.* Cleveland, Ohio, Lincoln Electric Co, eleventh edition 1956. Covers: processes, equipment, nomenclature, weldability of metals, techniques, procedures, speeds and costs for welding mild steel, design data for welded construction, designing arc welded structures, inspection and testing, applications.

' Machine design ': *Fastening and joining: reference issue,* Cleveland, Ohio, Penton Publishing Co, fourth edition 1967.

Oates, J A: *Welding engineers' handbook.* London, Newnes, second edition 1968. Covers a wide range of welding processes from manual metal arc welding to electroslag welding, and deals with a wide range of materials. Much information is presented in tabular form.

Phillips, A L: *Welding handbook.* New York, American Welding Society, five volumes. Vol 1 Fundamentals of welding, sixth edition 1968; 2 Welding processes: gas, arc and resistance, fifth edition 1964; 3 Special welding processes and cutting, fifth edition 1965; 4 Metals and their weldability, fifth edition 1966; 5 Application of welding, fifth edition 1967.

Skeist, I: *Handbook of adhesives.* New York, Reinhold, 1962. Information on the chemistry, manufacture and application of adhesives emphasising the selection of proper adhesives.

Soled, J: *Fasteners handbook.* New York, Reinhold, 1957. Comprehensive coverage of standard and proprietary fasteners from all manufacturers: rivets, inserts, screws, bolts, studs, nuts, washers, retaining rings, pins, nails, quick release fasteners, etc. Includes manufacturers directory.

FASTENERS see FABRICATION AND JOINING

Handbook and data books

Anderson, H H: *Centrifugal pumps*. Sutton, Surrey, Trade and Technical Press, 1962. Information and data on the design, construction and operation of centrifugal pumps.

British Compressed Air Society: *Technical reference book of compressed air terms and standards*. London, the Society, fifth edition 1961. Covers the installation, maintenance and utilisation of compressed air equipment and includes tables for use with compressed air and gases, rock drills and pneumatic tools, etc. Includes a directory and buyers' guide.

Compressed Air and Gas Institute: *Compressed air and gas handbook: a reference book on all phases of industrial air and gas compression and compressed-air powered portable tools and rock drills used by industry*. New York, the Institute, third edition 1961.

Compressed Gas Association Inc: *Handbook of compressed gases*. New York, Van Nostrand, 1966. Technical information on properties, uses, manufacture, safe handling and transportation of forty nine widely used gases.

Diels, K and Jaeckel, R: *Leybold vacuum handbook*. Translated from the German by H Adam and J Edwards. New York, Pergamon Press, 1966. Presents a vast amount of data relevant to the design of vacuum systems, with extensive tables covering the physical properties of the materials of vacuum technology.

Hydraulic handbook. Morden, Surrey, Trade and Technical Press, 1967. Data book on industrial hydraulics covering the design and manufacture of equipment which incorporates hydraulics. Section 1 components, systems and applications; 2 technical data 3 buyers' guide.

Karassik, I J: *Engineer's guide to centrifugal pumps*. New York, McGraw-Hill, 1964. Covers problems relating to the application, design, installation, operation and maintenance of centrifugal pumps.

' *Machine design*': *Fluid power: reference issue*. Cleveland, Ohio, Penton Publishing Co, third edition 1966. Design data, including basic information on fluid power systems, applications and selection of hydraulic and pneumatic power components. Includes manufacturers and products directory.

Pneumatic handbook. Morden, Surrey, Trade and Technical Press, 1966. Information, tables, charts and formulae on the design and utilisation of pneumatic equipment. Includes manufacturers and products directory.

Pumping manual. Compiled by the editors of *Pumps, pompes, pumpen.* Morden, Surrey, Trade and Technical Press, third edition 1968. Section 1 materials, systems, components, fittings; 2 pump performance, selection and operation; 3 technical data; 4 survey of pumps currently available; 5 manufacturers catalogue and buyers' guide.

Steinherz, H A: *Handbook of high vacuum engineering.* New York, Reinhold, 1963. Covers the design, construction, operation and maintenance of high vacuum systems and the major applications of high vacuum technology.

Streeter, V L: *Handbook of fluid dynamics.* New York, McGraw-Hill, 1961. Practical engineering coverage of fluid flow principles, data, methods and theory.

Tsiklis, D S: *Handbook of techniques in high pressure research and engineering.* Translated from the Russian by A Peabody. New York, Plenum Press, 1968. Covers a wide range of equipment and procedures used with high pressure.

Yeaple, F D: *Hydraulic and pneumatic power and control: design, performance, application.* New York, McGraw-Hill, 1966. A practical design handbook covering techniques for designing hydraulic and pneumatic systems for machines and processes.

Tables
Benedict, R P and Steltz, W J: *Handbook of generalised gas dynamics.* New York, Plenum Press, 1966. Generalised compressible flow tables for practising engineers who are confronted by problems in gas dynamics.

Jordan, D P and Mintz, M D: *Air tables: tables of the compressible flow functions for one-dimensional flow of a perfect gas and of real air.* New York, McGraw-Hill, 1965.

Directories
British Compressed Air Society: *Directory and buyers' guide for compressed air equipment.* London, the Society, 1966.

British Pump Manufacturers' Association. *British pumps: buyers' guide* 1966. The Association, 1966.

FORGING see WORKSHOP TECHNOLOGY

FOUNDRY PRACTICE
Dictionaries and encyclopedias
Schulenberg, A: *Giesserei lexikon.* Berlin, Fachverlag Schiele & Schon GmbH, fifth edition 1967. Encyclopedic dictionary of foundry practice.
Handbooks and data books
American Foundrymen's Society. *Cast metals handbook.* Des Plaines, Illinois, the Society, fourth edition 1957. Physical and mechanical properties, working and applications of: gray and white cast irons, malleable cast iron, nodular cast iron, steel castings, non-ferrous alloys.

American Foundrymen's Society: *Foundry sand handbook.* Chicago, the Society, sixth edition 1952. Procedures and standards for sand testing.

American Society for Metals: *Casting design handbook.* Novelty, Ohio, the Society, 1962. Pt 1 relations between the design of castings and processes for their production in the factory; 2 data on the properties of the principal cast metals and information on the selection and application of metals in cast form.

Investment Casting Institute: *Investment casting handbook.* Chicago, the Institute, 1969.

Great Britain. Ministry of Supply: *A handbook on die castings for the use of service designers and inspectors.* London, HMSO, 1953. Concise treatment of the advantages and limitations of processes.

Roll, F: *Handbuch der giesserei-technik.* Berlin, Springer-Verlag, two volumes in three 1959-1963.

Steel Founders Society of America: *Steel castings handbook.* Cleveland, Ohio, the Society, third edition 1960.
Directories
Durlanger, J: *Foundry directory and register of forges.* London, Standard Catalogue Co, 1967.

GAS TURBINES see ENGINE DESIGN AND APPLICATION

Dudley, D W: *Gear handbook: the design, manufacture and application of gears*. New York, McGraw-Hill, 1962. Covers most gears in commercial use; includes data on efficiency, loading practice, tolerances, inspection equipment, performance and life expectation of gears, gear arrangements.

Scoles, C A and Kirk, R: *Gear metrology*. London, McDonald, 1969. Methods of analytical measurements of pitch, profile, lead, etc are followed by details of commercial gear testing equipment.

HEAT TRANSFER see APPLIED HEAT

HEATING AND VENTILATING ENGINEERING
Handbooks and data books
American Society of Heating, Refrigerating and Air-conditioning Engineers: *ASHRAE guide and data book*. Two volumes, 1968. Comprehensive coverage of American practice. Volume 1 fundamentals and equipment, includes a manufacturers catalogue; volume 2 covers applications.

Brightside Heating and Engineering Co: *Technical data in the field of heating and air treatment engineering*. Sheffield, Brightside, 1964.

Carrier Air Conditioning Co: *Handbook of air conditioning system design*. New York, McGraw-Hill, 1965. Covers all important design steps from load estimation and air distribution to a breakdown of all-air, air-water, and water and direct expansion systems.

Emerick, R H: *Heating handbook: a manual of codes, standards and methods*. New York, McGraw-Hill, 1964. Guide to good practice covering the complete fields of heating, from storage of fuels to ultimate control of smoke and air pollution.

Fischer, L J: *Combustion engineers' handbook*. London, Newnes, 1961. Condensed English edition of the *Walther-Taschenbuch*, a German handbook covering combustion engineering, including data of particular value to engineers responsible for operating large boilers. Includes both British and German standards.

Porges, J: *Handbook of heating, ventilating and air conditioning*. London, Newnes, fifth edition 1964. Information, data, charts and tables for the practical heating engineer, reflecting British practice.

Strock, C and Koral, R L: *Handbook of air conditioning, heating and ventilating*. New York, Industrial Press, second edition 1965. Tables, formulae, graphs, maps and standard information for engineers. Includes climatic data (USA), load calculations, air conditioning systems, refrigeration for air conditioning, air handling and ventilation, fuels and combustion, space heating, piping, etc.

Directories
Air conditioning, heating and refrigeration news directory. Detroit, Business News Publishing Co, annual.

H & V R catalogue of heating and directory of heating engineers. Thornton Heath, Surrey, Heating and Ventilating Publications Ltd. Annual catalogue including specifications of plant and materials.

Heating and ventilating year book. London, Heating and Ventilating Contractors' Association. In addition to a buyers' guide, includes information on British and European standards.

INSTRUMENTATION see AUTOMATION AND INSTRUMENTATION

INTERNAL COMBUSTION ENGINES see ENGINE DESIGN AND APPLICATION

IRON AND STEEL
Encyclopedias and dictionaries
Osborne, A K and Wolstenholme, M J: *An encyclopaedia of the iron and steel industry*. London, Technical Press, second edition 1967. Concise descriptions of the materials, plant, tools and processes used in the iron and steel and related industries.

Handbooks and data books
Aitchison, L and Pumphrey, W I: *Engineering steels: a study of the properties of steels and the principles governing their selection for engineering applications*. London, McDonald, 1953.

British Steel Castings Research Association: *British and foreign specifications for steel castings*. Sheffield, the Association, third edition 1968. Presents in summary form the main requirements for steel castings of the national standardising organisations in various countries.

International Nickel Co: *Nickel alloy steels data book*. New York, the Company, third edition 1968. Summarises engineering properties and metallurgical characteristics of compositions that have been produced in commercially significant quantities.

217

'Machine design': *Ferrous metals book*. Cleveland, Ohio, Penton Publishing Co, 1961. Thirty data sheets, each covering the mechanical, physical and fabrication characteristics of a small group of similar alloys. Includes a products directory and manufacturers' guide.

Directories

Cordero, H G: *Iron and steel works of the world*. London, Metal Bulletin, fifth edition 1969. Arranged alphabetically by country, gives company details and information on plant and products. Includes classified buyers' guide.

Steel Founders' Society of America. *Directory of steel founders in the United States, Canada and Mexico*. Rocky River, Ohio, the Society, 1968.

LOCOMOTIVE ENGINEERING

Dictionaries and encyclopedias

Carter, E F: *The railway encyclopedia*. London, Starke, 1963. Information on the history, construction, engineering and methods of operation of the railways of the United Kingdom. Includes technical railway terminology.

Ransome-Wallis, P: *Concise encyclopaedia of world railway locomotives*. London, Hutchinson, 1959. A review of locomotive power covering the constructional details, problems of operation and the testing of locomotives.

Directories

Railway directory and yearbook. London, Transport and Technical Publications Ltd. World coverage of manufacturers of locomotive equipment, statistical information on production of locomotives, etc.

Sampson, H: *Jane's world railways*. London, Sampson Low, eleventh edition 1968. Tabulated data on world railways, railway systems and manufacturers; underground railways systems; track and signalling equipment; diesel engine manufacturers, etc.

LUBRICATION see TRIBOLOGY

MACHINE DESIGN

Handbooks and data books

Chironis, N P: *Machine devices and instrumentation*. New York, McGraw-Hill, 1966. A compilation of mechanical, electrical, hydraulic, pneumatic, optical, thermal and photographic devices for achieving a variety of functions and motions in automatic machines and instruments.

Ingenious mechanisms for designers and inventors. Four volumes. Vols 1-2 edited by F D Jones, 1930, 1936; vol 3 edited by L H Horton, 1951; vol 4 edited by J A Newell and L H Horton, 1967. New York, Industrial Press. Mechanisms and mechanical movements selected from automatic machines and various other forms of mechanical apparatus as outstanding examples of ingenious designs embodying ideas or principles applicable in designing machines or devices requiring automatic features or mechanical control.

Rothbart, H A: *Mechanical design and systems handbook.* New York, McGraw-Hill, 1964. New methods and operating data on design, performance and control of mechanical systems.

See also ENGINE DESIGN AND APPLICATION; FLUID POWER AND PNEUMATICS; GEARS; NOISE CONTROL; SPRINGS; VIBRATION.

MACHINE TOOLS see WORKSHOP TECHNOLOGY

MACHINING see WORKSHOP TECHNOLOGY

MARINE ENGINEERING

Dictionaries and encyclopedias

De Kerchove, R: *International maritime dictionary: an encyclopedic dictionary of useful maritime terms and phrases together with equivalents in French and German.* Princeton, NJ, Van Nostrand, second edition 1961.

Watson, G O: *Dictionary of marine engineering and nautical terms.* London, Newnes, 1965. Aims to define briefly terms likely to be met with in daily contact between personnel engaged in construction and operation of merchant ships.

Handbooks and data books

Great Britain. Admiralty. *Naval marine engineering practice.* London, Admiralty, second edition, two volumes 1959-1962. Descriptions and operating information on typical machinery, boilers and systems of the marine engineering division of the Admiralty.

American Bureau of Shipping: *Rules for building and classing steel vessels.* New York, annual.

Barr, H: *MacGibbon's MOT orals and marine engineering knowledge.* Glasgow, James Munro, twelfth edition 1962. Presented in the form of over 800 questions with answers given in language which will be readily understood by the practical engineer. Covers: strength of materials, ship construction, heat, definitions, electricity, turbines, steam engines, condensers and auxiliaries, refrigeration, boilers, oil fuel, internal combustion engines, tables and data.

Bowden, J K: *Sothern's marine diesel oil engines: a manual of marine oil engine practice.* Glasgow, James Munro, tenth edition 1968.

Fox, W J and McBirnie, S C: *Marine steam engines and turbines.* London, Newnes, second edition 1961.

Mackrow, C and Woollard, L: *Mackrow's naval architects and shipbuilders pocket book: formulae, rules and tables for marine engineers and surveyors.* London, Technical Press, fifteenth edition 1954.

'*Motor ship*' *reference book.* London, Temple Press, twenty first edition 1965. Aims to provide technical information on every type of marine diesel engine produced throughout the world.

Pounder, C C: *Marine diesel engines.* London, Newnes, fourth edition 1968. Covers fundamental theory, description of proprietary makes from eight countries, maintenance, automation, etc.

Rules and regulations for the construction and classification of steel ships, 1969. London, Lloyd's Register.

Ships and marine engines. Haarlem, H Stam NV, 1948-. Vol I theoretical shipbuilding; II resistance, propulsion and steering of ships; IIIA practical shipbuilding: steel ship construction; IIIB practical shipbuilding: rigging, equipment and outfitting of seagoing ships; IIIc shipbuilding and shipyard layout; IV design of merchant ships; V small seagoing craft and vessels for inland navigation; VI floating dredgers; VII marine engines and marine installation; VIII marine gears.

Sothern, J W M and Bowden, J K. *Verbal notes and sketches for marine engineer officers: a manual of marine steam engineering practice.* Glasgow, James Munro, nineteenth edition, two volumes 1958.

Directories

British shipbuilding compendium: the buyers' guide to British shipbuilding, marine engineering, repairing, ships and shipyard equipment. London, British Shipbuilding Compendium, thirteenth edition 1963.

Directory of shipowners, shipbuilders and marine engineers. London, Tothill Press, annual.

International shipping and shipbuilding directory. London, Benn, annual.

Lloyd's register of ships. London, Lloyd's Register. An annual containing names, classes and general information on ships classed by Lloyd's, and particulars of all known ocean-going merchant ships in the world of 168 tons and upwards.

MATERIALS HANDLING
Dictionaries and encyclopedias
Woodley, D R: *Encyclopaedia of materials handling*. Oxford, Pergamon Press, two volumes 1964. Covers various forms of equipment, systems of equipment selection and specialised divisions of handling such as unitisation, sheet and plate handling, etc.

Handbooks and data books
Bolz, H A and Hogemann, G E: *Materials handling handbook*. New York, Ronald Press, 1958. Covers: methods for analysing handling problems, principles, procedures and techniques for effective operation, systems design and installation, integration of materials handling with manufacturing processes, design, selection and classification of materials handling equipment.

Directories
'Mechanical handling': *Mechanical handling directory*. London, Iliffe, 1962. Classified guide to equipment including specifications and descriptions, trade names, lists of manufacturers.

MATERIALS SCIENCE AND ENGINEERING
Dictionaries and encyclopedias
Brady, G S: *Materials handbook: an encyclopedia for purchasing agents*. New York, McGraw-Hill, ninth edition 1963. Properties and characteristics of materials to assist in their selection for particular purposes. Scope includes, metals, alloys, refractories, abrasives, woods, synthetic resins, industrial chemicals, etc. Arranged alphabetically by material.

Handbooks and data books
Hetenyi, M: *Handbook of experimental stress analysis*. New York, Wiley, 1950. Eighteen chapters each dealing with either a principal

method, *eg* mechanical gauges, or a major topic of interest, *eg* residual stress.

Lipson, C: *Handbook of stress and strength: design and material applications.* London, Macmillan, 1963. Pt I consideration of stress; II consideration of strength; III balancing stress and strength; IV elastic deflection and elastic stability; V charts.

Mantell, C L: *Engineering materials handbook.* New York, McGraw-Hill, 1960. Reference data on design, structure and serviceability of engineering materials (metals, organic and inorganic), with emphasis on the fabricated forms of materials, their physical properties, adaptations, advantages and limitations. Forty three chapters each covering a particular class of material.

North Atlantic Treaty Organisation. Advisory Group for Aerospace Research and Development: *Handbook of brittle material design technology,* NATO, 1966. Summaries for the designer of data on the use of brittle non-metallic refractory materials for structural purposes.

Parker, E R: *Materials data book for engineers and scientists.* New York, McGraw-Hill, 1967. Engineering design data on solid materials, metals, ceramics, plastics, woods. Includes addresses of suppliers from which additonal information may be obtained.

Roark, R J: *Formulas for stress and strain.* New York, McGraw-Hill, fourth edition 1965. A summary of important formulae, facts and principles covering the strength of materials intended as a working handbook for engineers concerned with machine and structural design.

Sax, N I: *Dangerous properties of industrial materials.* New York, Reinhold, third edition 1968. Covers the hazards associated with the use of 12,000 common industrial and laboratory materials.

Touloukain, Y S: *Thermophysical properties of high temperature solid materials.* 6 vols in 9. New York, Macmillan, 1967. Results of a two year project carried out at the Thermophysical Properties Research Center, Purdue University. 8,500 pages of data covering the specific properties of elements, non-ferrous alloys, ferrous alloys, intermetallics, oxides, etc.

See also: NON-DESTRUCTIVE TESTING.

MECHANICAL HANDLING see MATERIALS HANDLING

Dictionaries and encyclopedias

Hall, N: *Dictionary of metal finishing chemicals.* Westwood, NJ, Metals and Plastics Publications, 1963.

Handbooks and data books

American Society for Metals: *Metals handbook. Vol 2: heat treating, cleaning and finishing.* Novelty, Ohio, the Society, eighth edition 1964. Detailed coverage of the equipment, control, applicability and results of processes for treating ferrous and non-ferrous metals.

Dettner, H W and others: *Handbuch der galvanotechnik. Vol 2: verfahren fur die galvanische und stromlose metallabscheidung.* Munich, Hauser, 1966. A comprehensive handbook on electroplating and metal finishing.

Enyedy, R: *Handbook of barrel finishing.* New York, Reinhold, 1965. Pt I finishing departments, layouts, components and equipment; 2 finishing methods including case histories.

Graham, A K: *Electroplating engineering handbook.* New York, Reinhold, second edition 1962. Engineering fundamentals and general processing data applicable to the electroplating and metal finishing industries.

Ollard, E A and Smith, E B: *Handbook of industrial electroplating.* London, Iliffe, third edition 1964. Covers: electrical equipment and deposition plant, comprehensive range of solutions and plating formulae, testing of deposits and solutions, etc.

Directories

Berrangé, J C: *Directory to the finishing trade.* London, Benn, 1968.

Finishing handbook and directory. London, Sawell Publications, annual. Includes encyclopedia of finishing, in addition to specialised chapters on electroplating processes and plastic powder spraying, etc, tables, British standards and a directory section.

Industrial finishing yearbook. London, Arrow Press. A technology section summarises the main trends in finishing processes, also includes tables, a list of official finishing specifications and British standards and a buyers' guide.

Metal finishing guide book and directory. Westwood, NJ, Metals and Plastics Publications Inc. An annual containing substantial technical information on: government finishing specifications, mechanical surface preparation, chemical surface preparation, plating solutions, control, special plating procedures, special surface treatments, organic finishing, finishing plant engineering, tables, directory.

METAL WORKING see WORKSHOP TECHNOLOGY

METALLURGY

Dictionaries and encyclopedias

Birchon, D: *Dictionary of metallurgy*. London, Newnes, 1965. British and American terms of interest to those concerned with the selection and application of metals. Emphasis on composition, physical properties and behaviour of materials.

Henderson, J G and Bates, J M: *Metallurgical dictionary*. New York, Reinhold, 1953. Definitions for physical and production metallurgy and related fields.

Merriman, A D: *A concise encyclopaedia of metallurgy*. London, McDoanald and Evans, 1965. A revision of Merriman's *Dictionary of metallurgy* originally published in 1958.

Rolfe, R T: *A dictionary of metallography*. London, Chapman and Hall, second edition 1949.

Handbooks and data books

American Society for Metals: *Metals handbook. Vol 1: properties and selection of materials*. Novelty, Ohio, the Society, eighth edition 1961. First of a multivolume set to supercede the single volume seventh edition. Vol 2, see METAL FINISHING; vol 3, see WORKSHOP TECHNOLOGY. Other volumes in preparation covering *Forming, forging and casting* and *Welding*. Contents of vol 1: Definitions and reference tables; carbon and low alloy steels, cast irons; stainless steels and heat resisting alloys; tool materials; magnetic, electrical and other special purpose materials; aluminium and aluminium alloys; copper and copper alloys; lead and lead alloys; magnesium and magnesium alloys; nickel; tin; titanium; zinc; precious metals; properties of pure metals.

American Society of Mechanical Engineers: *ASME handbook. 1: metals properties*. New York, McGraw-Hill, 1954. Tabulates the properties of metals (strength, hardness, machinability, composition, etc) about which the design engineer needs information.

American Society of Mechanical Engineers: *ASME handbook. 3:*

metal engineering—design. New York, McGraw-Hill, second edition 1965. Covers the characteristics of metals and their significance to engineers in the production of metal products.

Aerospace structural metals handbook. New York, Syracuse University Press, two volumes 1964. Vol I ferrous alloys; 2 non-ferrous alloys. Physical, chemical and mechanical property information on 138 metals and alloys of interest for aerospace structural applications. Loose-leaf format with periodic updatings.

'Machine design': *Metals: reference issue.* Cleveland, Ohio, Penton Publishing Co, 1970. Includes data on cast ferrous metals; wrought ferrous metals; non-ferrous metals; fabrication processes.

National Engineering Laboratory: *Creep data sheets.* London, Department of Scientific and Industrial Research, 1963. Summaries of data available to the creep information centre on high-temperature materials.

North Atlantic Treaty Organisation. Advisory Group for Aerospace Research and Development: *Material properties handbook.* NATO, four volumes 1966. Vol 1 aluminium alloys; 2 steels; 3 magnesium, nickel and titanium alloys; 4 superalloys. Information on the properties of aircraft materials produced in the various NATO countries presented in the form of data sheets in loose-leaf format to facilitate updating.

Pearson, W B: *A handbook of lattice spacings and structures of metals and alloys.* London, Pergamon Press, two volumes 1958, 1967. Presents in tabular form the total knowledge of the structures and lattice parameters of metals and alloy phases complete to the end of 1963.

Smithells, C J: *Metals reference book.* London, Butterworth, fourth edition, three volumes 1967. A survey of data relating to metallurgy and metal physics presented mainly in the form of tables and diagrams with a minimum of descriptive matter.

Steels' specifications handbook: cross index of chemically equivalent specifications and identification codes for ferrous and non-ferrous alloys. Cleveland, Ohio, Penton Publishing Co, 1953.

Woldman, N E: *Engineering alloys: names, properties, uses.* New York, Reinhold, fourth edition 1962. Lists the chemical composition, the mechanical and physical properties and the uses and general characteristics of 35,000 proprietary, commercial and technical alloys manufactured in the United States and abroad.

Directories

Metal bulletin handbook. London, Metal Bulletin Ltd, 1968. World statistics on the production of iron and steel, non-ferrous metals; ferrous alloys and iron and steel scrap.

Standard metal directory. New York, American Market Co. A biennial publication giving details of American iron and steel plant, metal smelters, rolling mills, etc.

See also: ALUMINIUM; COPPER; CORROSION; FOUNDRY PRACTICE; IRON AND STEEL; METAL FINISHING; NON-FERROUS METALS; POWDER METALLURGY; WIRE; WORKSHOP TECHNOLOGY.

METROLOGY

Handbooks and data books

American Society of Tool and Manufacturing Engineers: *Handbook of industrial metrology.* Englewood Cliffs, NJ, Prentice-Hall, 1967. Covers theoretical aspects of physical measurement in industry and applications of the various devices used.

Farago, F T: *Handbook of dimensional measurement.* New York, Industrial Press, 1968. Emphasis on measuring techniques and equipment which have been developed in recent years.

NOISE CONTROL

Harris, C M: *Handbook of noise control. New York,* McGraw-Hill, 1957. Covers the nature of noise, its measurement and techniques of noise control in buildings, industry, transportation and the community. Includes information on practical engineering techniques and reference data on such topics as problems of vibration and its control and the uses of various materials and methods in controlling industrial noise.

NON-DESTRUCTIVE TESTING

McMaster, R C: *Non-destructive testing handbook.* New York, Ronald Press, two volumes 1963. Covers all major methods of non-destructive testing, the detection of discontinuities, and the prediction or performance capabilities without impairing serviceability.

NON-FERROUS METALS
Dictionaries and encyclopedias
Simons, E N: *A dictionary of alloys*. London, Muller, 1960. Properties, compositions, functions and applications of a wide range of non-ferrous alloys.

Handbooks and data books
Hampel, C A: *Rare metals handbook*. New York, Reinhold, second edition 1961. Data on forty five of the less common metals—attempts to collate available reference data covering methods of production, chemical and physical properties, fabrication techniques and uses.

Pagonis, G A: *Light metals handbook*. Princeton, NJ, Van Nostrand, 1955. Data on the mechanical properties, physical properties, choice of alloy, casting, formability, machinability and joining methods for a variety of aluminium and magnesium base alloys.

Liddell, D M: *Handbook of non-ferrous metallurgy*. New York, McGraw-Hill, second edition 1945.

Simons, E N: *Guide to uncommon metals*. London, Muller, 1967. Practical reference data for engineers and metallurgists.

Directories
European metals directory. London, Quin Press, 1964. Guide to European producers and suppliers of non-ferrous metals, alloys, ores and semi-finished parts.

See also: ALUMINIUM; COPPER.

NUCLEAR ENGINEERING
Dictionaries and encyclopedias
Barnes, D E and others: *Newnes concise encyclopaedia of nuclear energy*. London, Newnes, 1962. Scope includes subjects from general technology to which nuclear engineers have cause to refer regularly.

Del Vecchio, A: *Concise dictionary of atomics*. New York, Philosophical Library, 1964. Non-technical definitions intelligible to the non-specialist, and brief biographies of important scientists and engineers.

Sarbacher, R I: *Encyclopedic dictionary of electronics and nuclear engineering*. New York, Prentice-Hall, 1959.

Handbooks and data books
Etherington, H: *Nuclear engineering handbook*. New York,

McGraw-Hill, 1958. Covers the industrial applications of nuclear energy.

Glasstone, S: *Sourcebook on atomic energy*. Princeton, NJ, Van Nostrand, third edition 1967. Brings together facts covering the past history, current status and possible future developments of atomic energy.

United States Atomic Energy Commission: *Reactor handbook*. New York, Interscience, second edition, four volumes in five 1960-1964. Vol 1 materials; 2 fuel processing; 3A physics; 3B shielding; 4 engineering.

Weinstein, R and others: *Nuclear engineering fundamentals*. New York, McGraw-Hill, 1964. Five books bound in one, covering atomic physics, nuclear physics interaction of radiation and matter, nuclear materials, and nuclear reactor theory.

Tables
Kunz, W and Schintlmeister, J: *Nuclear tables. First part: nuclear properties*. Two volumes 1963. *Part II: nuclear reactions*. Two volumes 1965, 1967. Oxford, Pergamon Press.

Directories
World nuclear directory: an international reference book. London, Harrap, third edition 1966. Arranged in alphabetical order of country, covers: national atomic energy agency, privately sponsored research organisations, university departments, professional societies, etc.

PIPING
Handbooks and data books
Crocker, S: *Piping handbook*. Fifth edition revised by R C King. New York, McGraw-Hill, 1967. Covers the design, specifications, fabrication, installation, testing and inspection of complete piping systems. Sections are included on material selection and nuclear, marine and sewage piping.

Heating, Piping and Air Conditioning Contractors' National Association: *Standard manual on pipe welding*. New York, the Association, second edition 1951. Covers welded piping design and installation practices.

Welded Steel Tube Institute Inc: *Handbook of welded steel tubing*.

228

Cleveland, the Institute, 1968. Data on the uses of welded carbon and alloy steel tubing and welded stainless steel tubing and pipe.

PLANT ENGINEERING

Handbooks and data books

Elonka, S M: *Plant operators' manual*. New York, McGraw-Hill, second edition 1965. Practical manual for the maintenance and operation of plant including data on boilers, pumps, turbines and diesels.

Morrow, L C: *Maintenance engineering handbook*. New York, McGraw-Hill, second edition 1966. Covers the responsibilities and procedures of maintenance in manufacturing industries. Includes the planning, installation, operating and administering of a plant maintenance programme, maintenance manuals, work simplification, reliability indexes and equipment.

' Power ' editorial staff: *Plant energy systems: energy system engineering*. New York, McGraw-Hill, 1967. Based on special reports published in *Power* magazine. Topics included are: pumps, compressing of air, fans, refrigeration, conduit heating, piping, valves, corrosion, mechanical seals, etc.

Staniar, W: *Plant engineering handbook*. New York, McGraw-Hill, second edition 1959. Covers all aspects of plant organisation, design, construction, operation and maintenance in the economic, mechanical, chemical and power areas of industrial plant operation. Much of the information is presented in the form of formulae, tables and graphs.

PLASTICS

Dictionaries and encyclopedias

Simonds, H R: *The encyclopedia of plastics equipment*. New York, Reinhold, 1964. Over 200 articles covering all types of processing equipment, extruders, calendars, moulding presses, etc, new processes and automatic operation.

Wordingham, J A and Reboul, P: *Dictionary of plastics*. London, Newnes, 1964. Concise guide to terms in everyday use. Appendices include trade names used in the plastics industry.

Handbooks and data books

' Machine design ': *Plastics: reference issue*. Cleveland, Ohio, Penton Publishing Co, 1968. Aims to present the design engineer with pertinent information and data necessary in the selection of plastics for particular design needs. Includes a manufacturers and products directory.

Oleesky, S S and Mohr, J G: *Handbook of reinforced plastics of the Society of the Plastics Industry, Inc.* New York, Reinhold, 1964. A useful sourcebook for calculating configurations for reinforced plastics; full information on moulding methods, tool design and properties of materials.

Simonds, H R: *Concise guide to plastics.* New York, Reinhold, second edition 1963. Information on the strength, properties, processes and production of plastics and their suitability for particular applications. Includes lists of manufacturers of plastics in the United States and trade names.

Society of the Plastics Industry: *SPI plastics engineering handbook.* New York, Reinhold, third edition 1960. A standard reference work on plastics materials, methods and fabrication. Covers accepted standards and specifications, methods of testing, processing, finishing and assembly techniques.

Wilson, F W: *Plastics testing and manufacturing handbook: a reference book on the use of plastics as engineering material for tool and workpiece fabrication.* Englewood Cliffs, NJ, Prentice-Hall for American Society of Tool and Manufacturing Engineers, 1965.

Directories

' *British plastics* ' *year book: a classified guide to the British plastics industry.* London, Iliffe.

Lawrence, J R: *Directory of the plastics industry.* Englewood Cliffs, NJ, Cahners Publishing Co.

' *Modern plastics* ' *encyclopedia issue.* Bristol, Connecticut, Plastics Catalog Corporation. An annual guide to: plastics, resins and moulding compounds, foamed plastics, chemicals and additives, film sheeting, laminates and reinforced plastics, machinery and methods, fabricating and finishing, buyers' guide and directory.

POWDER METALLURGY
Handbooks and data books

British Metal Sinterings Association: *BMSA specification manual.* Birmingham, the Association, 1964. Physical and mechanical properties and general comparisons of sintered metals.

Poster, A R: *Handbook of metal powders*. Princeton, NJ, Van Nostrand, 1966. Data on the purities, flow rate, bulk, density, etc of over 1,800 types of pure and alloyed powders. Includes data on metal powders used in refractory, nuclear, welding, aerospace, powder metallurgy and other industries.

Powder Metallurgy Equipment Association: *Powder metallurgy equipment manual. Vol 1 sintering furnaces and atmospheres, 1963. 2 compacting processes and tooling, 1965.* New York, the Association.

Directories

Precision Metal Molding Magazine: *Powder metallurgy directory.* Cleveland, Ohio. Annual.

POWER TRANSMISSION

Handbooks and data books

Association of Roller and Silent Chain Manufacturers: *Design manual for roller and silent chain drives.* The Association, 1955.

'Machine design': *Mechanical drives: reference issue.* Cleveland, Ohio, Penton Publishing Co, fourth edition 1969. Aims to provide the design engineer with information and data necessary to select and apply mechanical power transmission components and devices to meet his needs. Includes a manufacturers and products directory.

Williams, W A: *Mechanical power transmission manual.* New York, Conver-Mast, 1953. Covers the fundamentals of acceleration, power and work governing the selection of power transmission equipment, emphasising the basic principles underlying each type of drive.

PRODUCTION ENGINEERING

Dictionaries and encyclopedias

Encyclopedic dictionary of production and production control. Englewood Cliffs, NJ, Prentice-Hall, 1964. Definitions covering automatic control, budgeting, equipment, inventory, machinery and tooling, manufacturing processes, materials and materials handling, plant layout, product design, quality control, etc.

Handbooks and data books

American Society of Tool and Manufacturing Engineers: *Value engineering in manufacturing: a reference book on the theory, principles, applications and administration of value engineering and analy-*

231

sis in industry. Englewood Cliffs, NJ, Prentice-Hall, 1967. Includes principles of value engineering, role of management, value engineering techniques, organisation and staffing, training for value engineering.

Bolz, R W: *Production processes: the producability handbook.* Cleveland, Ohio, Penton Publishing Co, 1963. A broad survey of the processes utilized in the manufacture of all types of products and components.

Carson, G B: *Production handbook.* New York, Ronald Press, second edition 1959. Attempts to cover every phase and function of the planning, operation and control of present day industry.

Maynard, H B: *Industrial engineering handbook.* New York, McGraw-Hill, second edition 1963. Covers industrial engineering functions and methods; work measurement techniques; wage and salary plans; control procedures; plant facilities; mathematical and statistical procedures.

Wilson, F W and Harvey, P D: *Manufacturing, planning and estimating handbook: a comprehensive work on the techniques for analysing the methods of manufacturing a product and estimating its manufacturing cost.* New York, McGraw-Hill, 1963.

Tables

Eilon, S: *Industrial engineering tables.* London, Van Nostrand, 1962. Distribution functions; numerical tables; production and inventory control; work measurement; quality control.

See also AUTOMATION AND INSTRUMENTATION; FABRICATION AND JOINING; FOUNDRY PRACTICE; METAL FINISHING; QUALITY CONTROL; WORKSHOP TECHNOLOGY.

PUMPS see FLUID POWER AND PNEUMATICS

QUALITY CONTROL
Handbooks and data books
Ireson, W G and others: *Reliability handbook.* New York, McGraw-Hill, 1966. Summarises the state of the art of reliability engineering.

Juran, J M: *Quality control handbook.* New York, McGraw-Hill, second edition 1962. Covers economics of quality, specification of quality, organisation for quality; acceptance of control and assurance of quality; statistical methods of quality control and quality policy objectives.

REFRIGERATION ENGINEERING
Handbooks and data books

American Society of Heating, Refrigerating and Air Conditioning Engineers Inc: *ASHRAE guide and data book*. New York, the Society, 1968.

Busby, H D: *Refrigeration reference notebook*. Des Plaines, Illinois, Nickerson & Collins Co, 1968. Information presented mainly in tabular form, covering such subjects as temperature, pressure relationships of refrigerants, fittings and heat and costing loads.

Fidler, J C: *Manual of refrigeration practice*. London, Technical Productions Ltd, 1965. A design manual for the development of equipment for specific applications. Scope includes production and control of cold, application of cold, refrigeration in food technology and biology, industrial uses of refrigeration.

Woolrich, W R: *Handbook of refrigerating engineering*. Westpoint, Connecticut, Avi Publishing Co, fourth edition 1965.

Directories

Refrigeration and air conditioning directory. London, Refrigeration Press. An annual buyers' guide to the British industry.

SEALS

' Machine design ': *Seals: reference issue*. Cleveland, Ohio, Penton Publishing Co, fourth edition 1969. A handbook on seals, packaging and gaskets designed to provide the engineer with practical information on everyday sealing problems. Includes manufacturers and products directory.

SHIPBUILDING see MARINE ENGINEERING

SPRINGS

Gayer, J D and Stone, P H: *Helical spring tables: an easy to use index of ready designed compression and tension springs from which selections may be made with minimum calculations*. New York. Industrial Press, 1955.

STEAM ENGINEERING
Handbooks and data books

Spring, H M: *Boiler operators' guide*. New York, McGraw-Hill, 1940. Practical manual of steam boiler operation and maintenance. Covers: design, construction, characteristics, installation, etc.

8*

Tables

Electrical Research Association: *1967 steam tables: thermodynamic properties of water and steam, viscosity of water steam, thermal conductivity of water and steam*. London, Edward Arnold, 1967

Keenan, J H and Keyes, F G: *Steam tables—English units: thermodynamic properties of water, including vapor, liquid and solid phases*. New York, Wiley, second edition 1969.

National Engineering Laboratory: *Steam tables 1964: physical properties of water and steam, 0-800°C, 0-1000 bars*. London, HMSO, 1964.

Schmidt, E: *VDI-wasserdampftafeln mit einem Mollier (i,s)—Diagram bis 800°C. VDI steam tables including a Mollier (i,s)—diagram for temperatures up to 800°C*. Berlin, Springer Verlag, fourth edition 1956.

See also APPLIED HEAT; MARINE ENGINEERING

STEEL see IRON AND STEEL

STRENGTH OF MATERIALS see MATERIALS SCIENCE AND ENGINEERING

THERMODYNAMICS see APPLIED HEAT

TOOLING see WORKSHOP TECHNOLOGY

TRIBOLOGY

Handbooks and data books

Ellis, E J: *Fundamentals of lubrication*. London, Scientific Publications Ltd, 1968. Practical data on lubricants and their selection, prescription, specification, application and testing.

Institute of Petroleum: *IP standards for petroleum and its products*. Four volumes *Pt 1 Methods for analysis and testing*. twenty fifth edition 1966; *2 Methods for rating fuels—engine tests*. Second edition 1960; *3 Methods for assessing performance of crankcase lubricating oils—engine tests*. 1961; *4 Methods for sampling*. Second edition 1962. London, the Institute.

Lipson, C: *Handbook of mechanical wear: wear, frettage, pitting, cavitation, corrosion*. Ann Arbor, Michigan, University of Michigan Press, 1961.

MacGregor, C W: *Handbook of analytical design for wear*. New

York, Plenum Press, 1964. Data to enable the mechanical design engineer to select analytically those design parameters necessary to establish suitable wear characteristics of a mechanism.

O'Connor, J J and Boyd, J: *Standard handbook of lubrication engineering*. New York, McGraw-Hill, 1968. After a preliminary section on friction and its laws, individual chapters deal with lubrication of sliding and rolling bearings, gears, chains, engines, motors, compressors, etc. Lubrication in specific industries such as nuclear power plants is also covered.

VACUUM TECHNIQUES see FLUID POWER AND PNEUMATICS

VALVES

British Valve Manufacturers Association: *Technical reference book on valves for the control of fluids*. London, the Association, 1964.

VIBRATION

Handbooks and data books

Harris, C M and Crede, C E: *Shock and vibration handbook*. New York, McGraw-Hill, three volumes 1961. Vol 1 basic theory and measurement; 2 data analysis, testing and methods of control; 3 engineering design and environmental conditions. Covers all major areas of shock and vibration technology.

Nestorides, E J: *Handbook of torsional vibration*. Cambridge University Press, 1958. Presents data derived from research at the British Internal Combustion Engine Research Association. Covers formulae, design procedures, methods and specifications used in torsional vibration problems.

Tables

Bishop, R E D and Johnson, D C: *Vibration analysis tables*. Cambridge University Press, 1956.

WEAR see TRIBOLOGY

WELDING see FABRICATION AND JOINING

WIRE

Handbooks and data books

Dore, A B: *Steel wire handbook*. Stamford, Connecticut, Wire Association, 1965. Covers: production if wire rod, cleaning and coating in preparation for drawing, theory of wire drawing, wire drawing dies, lubrication for wire drawing.

Directories

' *Wire and wire products* ' *buyers' guide and yearbook of the Wire Association*. Stamford, Connecticut, Quinn-Brown Publishing Corp.

' *Wire industry* ' *yearbook*. London, Wire Industry Ltd. Includes buyers' guide, wire reference data, trade terms and processes, trade names.

WORKSHOP TECHNOLOGY

Dictionaries and encyclopedias

Clauser, H R and others: *The encyclopedia of engineering materials and processes*. New York, Reinhold, 1963. A work produced for technical staff in the manufacturing industries, consisting of 300 articles, each summarising either a group of materials or processes used in modern industry.

Winfield, H: *A glossary of metalworking terms*. London, Blackie, 1960.

Handbooks and data books

American Society for Metals: *Metals handbook. Vol 3: machining*. Novelty, Ohio, the Society, eighth edition 1967. Comprehensive coverage of all machining processes. Case histories give information on methods used in specific applications.

American Society of Mechanical Engineers: *ASME handbook. 4: metals engineering—processes*. New York, McGraw-Hill, 1958. Detailed data on the various processes by which metals are converted into finished products. Includes: heat treatment of steel, casting, hot working, cold working, powder metallurgy, welding and cutting, machining, finishing, electroforming.

Anderson, E P: *Audel's millwrights and mechanics guide for plant maintainers, builders, riggers, erectors, operators, construction men and engineers*. New York, Audel, fourth edition 1963. Intended as a practical guide to the problems of maintenance of mechanical machines.

Grant, H E: *Jigs and fixtures: non-standard clamping devices*. New York, McGraw-Hill, 1967. A reference book, international in scope, intended for the designers of production equipment covering over 1,600 clamping devices.

Habicht, F H: *Modern machine tools*. Princeton, NJ, Van Nostrand, 1963. A descriptive handbook giving an introduction to each major type of standard machine tool, and information on general construction, principles of operation and range of capabilities.

Jensen, J E: *Forging industry handbook*. Cleveland, Ohio, Forging Industry Association, 1966. Describes forging processes and methods, forging design principles, materials for forging, heat treatment, forging facilities and forging quality.

LeGrand, R: *New American machinists' handbook*. New York, McGraw-Hill, 1955. A quick reference aid for machinists, designers, draughtsmen and engineers, including tables and formulae covering machining, metal forming, assembly, materials, metal finishing, inspection, fastening devices, tool engineering, machine tool standards, and power transmission.

Machine tool specification manual: comparative specifications for the main classes of chipforming machine tools. London, MacLean Hunter, 1963.

Machinery's screw thread handbook. London, Machinery Publishing Co, nineteenth edition, 1965. Covers British, American and continental forms and series.

Metal Cutting Tool Institute: *Metal cutting tool handbook*. New York, the Institute, 1965. Data on twist drills, reamers, counterbores, taps, dies, milling cutters, gear shaper cutters, broaches, etc.

Metcut Research Associates Inc: *Machining data handbook*. Cincinnati, Metcut, 1966. Condensed information on the various machining operations used in the manufacture of army materials and products.

Wilson, F W: *Die design handbook*. New York, McGraw-Hill for American Society of Tool and Manufacturing Engineers, second edition 1965. Includes numerous practical designs for all types of die used in cold presswork practice, in addition to information on design of sheet metal products, die sets and elements, presses and assessories and materials.

Wilson, F W: *Handbook of fixture design*. New York, McGraw-Hill for American Society of Tool and Manufacturing Engineers, 1962. Covers design principles and working experience of jigs and fixtures for milling, drilling, boring, turning, broaching, grinding, etc.

Wilson, F W: *Plastics tooling and manufacturing handbook: a reference book on the use of plastics as engineering materials for tool and workpiece fabrication.* Englewood Cliffs, NJ, Prentice-Hall for American Society of Tool and Manufacturing Engineers, 1965.

Wilson, F W: *Tool engineers handbook.* New York, McGraw-Hill for American Society of Tool and Manufacturing Engineers, second edition 1959. A reference work covering all aspects of planning, control, design, tooling, and other operations involved in the mechanical processing of finished products.

Tables

Smith, W S: *Quadrant coordinate tables for engineering designers, jig and tool draughtsmen, programmers of automatic machine tools, markers off, and others whose work involves the calculation of run-out distances and circular or eliptical work of a similar kind.* London, Technical Press, 1967.

Directories

American Metal Stamping Association: *Guide.* Shaker Heights, Ohio, the Association. A periodical listing of members with details of facilities and services.

Directory of forging, stamping and heat treating plants. Pittsburgh, Steel Publication Inc, biennial.

Directory of metalworking machinery. Washington, US Government Printing Office, two volumes. Provides details of US metalworking manufacturers.

Durlanger, J: *Foundry directory and register of forges.* London, Standard Catalogue Co, 1967.

Machine and tool directory. Wheaton, Illinois, Hitchcock Publishing Co. An annual listing of manufacturers and products.

Machine tool directory. London, Mills and Robinson, 1964. An alphabetical list of firms with a classified index.

Metal working directory. New York, Dun and Bradstreet. An annual directory of metalworking plants in the USA.

See also FABRICATION AND JOINING; FOUNDRY PRACTICE; IRON AND STEEL; METAL FINISHING; PIPING; PLANT ENGINEERING; PRODUCTION ENGINEERING.

CHAPTER 13

BIBLIOGRAPHICAL CONTROL:
CURRENT BOOK SELECTION AND SUBJECT BIBLIOGRAPHY

BIBLIOGRAPHICAL CONTROL: The present and the following chapters are concerned with the bibliographical control of the literature of mechanical engineering. Bibliographical control refers to the processes of identifying and listing the various forms of document which cover a particular subject. Bibliographical control helps to obviate the duplication of effort by drawing the engineer's attention to documents which describe what work has already been carried out in a particular field. No research project or investigation which involves the expenditure of large amounts of money should ever be undertaken without a preliminary literature search.

The tools which effect bibliographical control are the various lists of new books, current and retrospective bibliographies, library catalogues, and abstracting, indexing and reviewing services. It will now be realised that some bibliographical tools covering specific forms of document, such as patent specifications and technical reports, have already been listed in previous chapters. This present chapter will cover book selection aids and subject bibliography, while chapter fourteen will concentrate on abstracting, indexing and reviewing services.

As pointed out in the introduction, this work is not intended as a bibliography of mechanical engineering, but rather as a guide to the literature of the subject; consequently no attempt has been made to list examples of individual textbooks or monographs. The rate of book publication is such that any attempt at systematic listing of individual titles would have rendered the work out-of-date even before publication. The approach has been to indicate to the engineer or information worker the tools with which he can undertake this selection for himself on a continuous basis.

GENERAL AIDS TO BOOK SELECTION

Every book published each week in the United Kingdom is listed in the *British national bibliography* (BNB), which has been published by the Council of the BNB since 1950. Complete bibliographical details are given for each title in a main classified sequence, arranged by the Dewey decimal classification, and this is supplemented by an author and title index, and for the last issue of each month a subject index. Cumulated issues of BNB are published each quarter and an annual volume is also available. Cumulated subject catalogues and indexes have also been published covering five-yearly periods since 1950. The books published each week in the United Kingdom through conventional book trade channels are also listed alphabetically by their authors and titles in *The bookseller,* the official organ of the book trade. These weekly lists cumulate, and the titles included then appear in *Whitaker's books of the month and books to come* and its classified counterpart the monthly *Current literature. Whitaker's cumulative booklist* is a quarterly publication covering books published during the periods January-March; January-July; January-September; and finally the complete year. Books are entered under classified groupings and also in an alphabetical sequence under their authors and titles. Publication since 1970 is in a separate supplement instead of in the journal.

The British Council's *British book news* is a monthly guide to book selection which presents descriptive annotations for a selection of recommended books, including those on applied science and engineering, published in the British Commonwealth. *British books in print . . . the reference catalogue of current literature* is an annual record, published by Whitaker in October of each year, of books on sale in the United Kingdom at the end of the previous April. Entries for each book included are given under both author and title.

American book publishing record (ABPR) is a monthly, published by Bowker, which cumulates, under a Dewey decimal classification subject arrangement, references to all books listed during the current month in the ' Weekly record ' section of *Publishers' weekly,* a journal which aims to provide bibliographical details of new American books as they are published. All references to books listed in the previous twelve monthly issues of ABPR are cumulated each year in the *American book publishing record cumulative.* In addition to covering books published in the United States the *Cumulative book index* (CBI) (H W Wilson, 1928-) is a monthly world list of books published in the English language which cumulates quarterly, annually, etc. Entries are arranged in one alphabetical sequence under author, title and subject. A major current subject bibliography is the *Library of Congress catalog. Books: subjects,* a quarterly alphabetical subject catalogue with annual and quinquennial cumulations which has been published since 1950. This catalogue includes publications on all subjects and in all languages as they are acquired by the Library of Congress and a number of cooperating libraries and thus it constitutes a record of a large proportion of the world's significant books. Headings are provided for MECHANICAL ENGINEERING and related subjects such as HEAT ENGINEERING, MACHINERY, MECHANICAL MOVEMENTS, POWER TRANSMISSION, PRODUCTION ENGINEERING, etc.

Subject guide to books in print: an index to the 'Publishers' trade list annual' (Bowker) is an annual guide, published since 1957, to the volumes available through the book trade in the United States. This publication adopts the same subject headings as those used in the above catalogue. National and book trade bibliographies published in other countries can be traced by referring to Collison, R L: *Bibliographies: subject and national.* London, Crosby Lockwood, third edition 1968.

AIDS TO THE SELECTION OF NEW TECHNICAL BOOKS
Aslib book list: a monthly list of recommended scientific and technical books with annotations has been published by Aslib since 1935. The list is intended as a selection tool for librarians, and includes brief descriptive annotations of an average length of fifty words. Arrangement is classified according to the Universal Decimal Classification and each item is graded into one of the following categories: A—an introductory text; B—intermediate standard textbook; c—advanced level text; D—directory or similar type reference work. Entries are limited almost exclusively to English language works and each issue contains about fifty annotations. A more extensive but similar publication is New York Public Library's *New technical books: a selective list with descriptive annotations.* This monthly publication lists mainly English language imprints according to the Dewey classification; annotations usually consist of a full contents list with a note indicative of treatment, level, etc. Each issue averages about 120 titles.

A useful evaluative selection guide is the Special Libraries Associa-

241

tion's monthly *Technical book review index* (TBRI), which arranges a selection of current technical books in alphabetical order of author and gives, for each, extracts from a recent review or reviews published in authoritative journals. Approximately 100 titles are reviewed in each issue. A very real disadvantage of this guide is its lack of a subject index. The lists of accessions issued by important libraries such as the Science Museum Library's monthly *List of accessions to the library* and the Ministry of Technology's monthly *Book list* can be used as additional checklists for new material. The London publishers Whitaker issue *Technical and scientific books in print* a useful annual reference catalogue, classified into 45 main groups with 300 sub-groupings, to technical books currently on sale in the United Kingdom. The 1969 edition listed over 25,000 volumes.

Book reviews or brief annotations of new publications are to be found in many technical periodicals. *Chartered mechanical engineer* and *Mechanical engineering* include in each monthly issue brief annotations of titles received into the Institution of Mechanical Engineers' library and the Engineering Societies' Library respectively. Examples of journals including more analytical or evaluative reviews are *Journal of applied mechanics* and *Materials science and engineering*.

RETROSPECTIVE BIBLIOGRAPHIES AND CATALOGUES OF TECHNICAL BOOKS
The most comprehensive catalogue of the world's literature on engineering is the *Classed subject catalog* of the Engineering Societies' Library published by Hall of Detroit in 1963. This catalogue, which covers all aspects of engineering and related sciences, is arranged by the Universal Decimal Classification and its 13 volumes (12 volumes of subject catalogue, 1 of subject index) reproduce 239,000 catalogue cards, for books, pamphlets, reports etc in all languages, from the library's card catalogue. Supplements to the main work are published periodically. A much more selective catalogue of 6,000 English and foreign language books published from 1930 to 1954 covering mechanical, civil, electrical, marine, sanitary and other branches of engineering but excluding metallurgy is the Science Museum Library's *Books on engineering* (HMSO, 1957). This catalogue is also classified by UDC

and it was published to provide the postal borrower with 'a list of books on the subjects most in demand'. H K Lewis and Co Ltd have offered a subscription library service since the ninteenth century, and their *Catalogue of Lewis' medical, scientific and technical lending library*, London, two volumes 1965-66 lists books acquired up to December 1963. Volume 1 is the author/title catalogue while volume 2 is a classified index of subjects. The 33,000 titles included are limited to English language editions, and supplements are issued at bi-monthly intervals.

A general guide to 8,000 books covering all aspects of science and technology is the *McGraw-Hill basic bibliography of science and technology* (New York, McGraw-Hill, 1966), which lists the titles under the 7,400 subject headings used in the *McGraw-Hill encyclopedia of science and technology* of which this work is a supplement. This bibliography is intended primarily for the science-oriented student and the interested layman, and each title is given a brief annotation indicative of level and treatment.

Important British technical books published during the period 1935-1957 can be traced by consulting Aslib's two volumes of *British scientific and technical books: a select list of recommended books published in Great Britain*, 1935-1952 and 1953-1957, published by James Clarke in 1956 and 1960 respectively. The volumes included are in the main selected from those listed over the years in *Aslib booklist;* arrangement is classified but while the A, B, C, D classification of the booklist is retained no annotations are included.

Coverage of American technical books published up to 1956 is given by Hawkins, R R: *Scientific, medical and technical books published in the United States of America*, Washington, National Research Council, second edition 1958. Inside a classified arrangement the contents and a brief annotation are given for 8,000 titles. A series of volumes covering technical literature from 1960 to 1965 were edited by P B Steckler for Bowker. These volumes, *American scientific books: a selection of scientific, technical and medical books as entered in the American book publishing record* were published for the periods 1960-1962; 1962-1963; 1963-1964 and 1964-1965. The arrangement of the volumes and the information given for each title is consistent with the arrangement of Hawkins. As no further issues of Steckler were published after 1965, subject selection of American technical books published since this date must now be undertaken through reference to the *Library of Congress catalog. Books: subjects* or the annual *Subject guide to books in print.*

Bibliographies of scientific subjects which list monographs, documents and periodical articles covering the subject can be traced by using bibliographies of bibliographies, the most comprehensive of which is Theodore Besterman's *World bibliography of bibliographies.* Lausanne, Societas Bibliographica, fourth edition, five volumes 1965-1966. This work lists only separately published bibliographies published up to the end of 1963 under a series of alphabetical subject headings. Bibliographies, both separately published and those included with other works, published since this time can be traced through *Bibliographic index: a cumulative bibliography of bibliographies,* which has been issued at six monthly intervals, with two year accumulations, by H W Wilson since 1942. Arrangement is again by subject under alphabetically arranged headings. The most comprehensive listing of bibliographies covering mechanical engineering is, however, given in volume 1 of the Engineering Societies' Library *Classed subject catalog.* In all, over 12,000 compilations covering all aspects of engineering up to 1961 are covered, including both separately published bibliographies and those appended to other works, in the UDC classified sequence. Additional subject access to these documents is given in volume 13, the alphabetical subject index to the classed catalogue, under the name of each specific subject with the subdivision—bibliography.

The Science Museum Library *Bibliographical series* commenced in 1931. Each issue, which has been compiled as a result of a subject request for information in the library, lists the articles and monographs published on a particular topic in science or technology. Although some 800 bibliographies have been published since 1931, the current rate of production is down to about one per year. Lists of relevance to the scope of this guide include *751: Marine corrosion of metals; 779: Reclaiming of used lubricating oil* and *771: Lapping and superfinishing.* A similar series of bibliographies has been issued by the Engineering Societies' Library in the *ESL bibliography* series. Each bibliography comprises an annotated list of books, pamphlets and periodical articles and compilations in the series include: no *3: Bibliography on precision instrument casting by the lost wax process*, 1949; *6: Bibliography on non-metallic bearings*, 1950; *11: Bibliography on machinery foundations: design, construction, vibration elimination*, 1955; *12: Bibliography on adhesive bonding of metals*, 1957 and *13: Bibliography on shell moulding*, 1959.

Few specific bibliographies of books on mechanical engineering have been published. Those which are currently available are limited to a general *Readers' guide to books on mechanical engineering* published by the County Libraries Section of the Library Association in 1958, and James N Siddall's very much more specific *Mechanical design: reference sources*, University of Toronto Press, 1967. Siddall's book is a convenient bibliography of over 4,500 brief references, without annotation, to textbooks, monographs and basic papers on such subjects as fasteners, gears, tool design etc. Within each subject category the references are listed chronologically by date of publication. Two other vitally important bibliographical tools for the mechanical engineer are the cumulated catalogues of papers published since their inception by the two major professional bodies:

Institution of Mechanical Engineers: *Brief subject and author index of papers published in the proceedings, 1847-1950*. 1951. An additional *Brief subject and author index of papers published by the Institution, 1951-1964* has been issued and this is now supplemented by an annual subject and author index covering all papers published in the *Proceedings, Journal of mechanical engineering science* and *Journal of strain analysis*.

and:

American Society of Mechanical Engineers: *Seventy-seven year index: technical papers, 1880-1956*. Easton, Pennsylvania, 1957. An author and subject index. An annual author and subject index to the papers published in *Mechanical engineering* and the constituent volumes of *ASME transactions* and also to *ASME miscellaneous papers* not published in these journals is issued as the *Society records*.

The following pages present a selection of subject bibliographies covering various aspects of mechanical engineering, and including all forms of document, which have been published since 1950.

ADHESIVES see FABRICATION AND JOINING

AUTOMATION AND INSTRUMENTATION
Bayley, F J and Turner, A B: *Bibliography of heat transfer instrumentation*. London, Treasury, Advisory Group for Aeronautics, 1968. (Reports & memoranda 3512.)

Brombacher, W G *et al*: *Guide to instrumentation literature*. Washington, US Government Printing Office, 1955. (US National Bureau of Standards circular 567.)

Hampshire Technical Research Industrial Commercial Service: *Numerical control of machine tools: a bibliography*. Southampton, Central Library, 1970.

US Department of Commerce. National Bureau of Standards : *Numerical controls: a review of selected US government research and development reports*. OTR-116, 1965.

BEARINGS
Geary, P J: *Knife-edge bearings: a bibliographical survey*. Chiselhurst, British Scientific Instruments Research Association, 1955.

Peters, A and Devlin, P: *Bibliography on gas lubricated bearings*. United States Atomic Energy Commission, 1965. (TID-22477.)

ENGINE DESIGN AND APPLICATION
American Society of Mechanical Engineers: *Bibliography on gas turbines, 1896-1948*. New York, the Society, 1962.

FABRICATION AND JOINING
American Welding Society: *AWS bibliographies, 1937-1961; compiled by E A Fenton*. New York, the Society, 1962.

Murphy, J: *Adhesive bonding: a select bibliography*. Hatfield, Hertfordshire Technical Library and Information Service, 1968.

Resistance Welder Manufacturers' Association: *Bibliography on resistance welding*. Philadelphia, the Association, 1953.

United Kingdom Atomic Energy Authority: *Atomic weapons research establishment. Ultrasonic welding, 1960-1965*. Aldermaston, AWRE, 1966.

National Aeronautics and Space Administration: Technical Utilization Division: *Bibliography on welding methods.* NASA, 1966. (NASA SP-5024.)

FLUID POWER AND PNEUMATICS

Brock, T E: *Fluidics applications: analysis of the literature and bibliography.* Cranfield, British Hydromechanics Research Association, 1968.

National Fluid Power Association: *Bibliography of fluid power.* Evanston, Illinois, the Association, 1957.

United Kingdom Atomic Energy Authority. Development and Engineering Group R & D Branch: *Bibliography on axial flow compressors, 1934-1958.* London, HMSO, 1960.

FOUNDRY PRACTICE

Lakner, J F: *Selected bibliography on hydrostatic fluid to fluid extrusion of metals.* United States Atomic Energy Commission, 1967. (UCID-15236.)

Rabe, R: *Bibliography of shell moulding, 1945-1960.* Des Plaines, American Foundrymen's Society, 1960.

GAS TURBINES see ENGINE DESIGN AND APPLICATION

HEATING AND VENTILATING ENGINEERING

National College for Heating, Ventilating, Refrigerating and Fan Engineering: *Fan engineering bibliography.* London, the College, 1965.

INSTRUMENTATION see AUTOMATION AND INSTRUMENTATION

IRON AND STEEL

Iron and Steel Institute: *Bibliographical series.* London, the Institute, 1930-. Each volume within the series consists of a comprehensive bibliography with abstracts covering a particular aspect of the properties and manufacture of iron and steel and is prepared by the institute's library staff. Recent volumes include *21b: Continuous casting of steel, 1959-1964; 23a: Ironfounding in the blast furnace, 1962-1965;* and *24: Vacuum metallurgy of steel, 1941-1965.*

LUBRICATION see TRIBOLOGY

METAL WORKING see WORKSHOP TECHNOLOGY

METALLURGY

North Atlantic Treaty Organisation. Advisory Group for Aerospace Research and Development: *A bibliography of refractory metals* (AGARD bibliography no 5); compiled by R Syre and K J Spence. Maidenhead, Technivision Services, 1968.

NON-DESTRUCTIVE TESTING

United Kingdom Atomic Energy Authority. Non-Destructive Testing Centre, Harwell: *Selected bibliographic guide to conference papers on non-destructive testing, 1955-1967.* Harwell, UKAEA, 1968.

NUCLEAR ENGINEERING

International Atomic Energy Agency: *Bibliographical series.* Vienna, the Agency, 1960-. Each volume issued within the series is a comprehensive bibliography with abstracts covering a peaceful application of atomic energy. Subjects covered include: *2: Nuclear reactors,* 1960; *3: Nuclear propulsion,* 1962, and *34: Recurring inspection of nuclear reactor steel pressure vessels,* 1968.

POWDER METALLURGY

Goetzel, C G: *Treatise on powder metallurgy.* New York, Interscience, 1963. Volume 4: classified and annotated bibliography.

PRODUCTION ENGINEERING

Hatfield Technical College Library: *Value engineering bibliography.* Hatfield, 1966.

QUALITY CONTROL AND MEASUREMENT

Mort, G: *Quality control.* London, Library Association, 1967.

SEALS

British Hydromechanics Research Association: *Bibliography on fluid sealing.* Harlow, the Association, 1962.

SPRINGS

Associated Spring Corporation: *Springs: a bibliography 1678-1956.* Bristol, Conn, the Corporation, 1957.

TRIBOLOGY

National Aeronautics and Space Administration. Technical Utilisation Division: *Bibliography on solid lubricants.* NASA, 1966. (NASA SP-5037.)

WELDING see FABRICATION AND JOINING

WIRE

British Iron and Steel Research Association. Development and information Services: *Bibliography on the production of steel wire.* London, the Assoc., 1966.

International Deep Drawing Research Group: *Bibliography on deep drawing, 1910-1958.* London, British Iron and Steel Research Association, 1959.

WORKSHOP TECHNOLOGY

American Society of Tool and Manufacturing Engineers: *Metal cutting bibliography, 1943-1956.* Detroit, the Society, 1960.

Battelle Memorial Institute: *DMIC report 179: a guide to the literature on high velocity metal working, 1879-1962.* Columbus, Ohio, the Institute, 1962.

Bickle, W H: *Crushing and grinding: a bibliography, 1867-1957.* London, HMSO, 1958.

Industrial Diamond Information Bureau: *Diamond as a cutting tool for metals and non-metallic materials, 1939-1961.* London, the Bureau, third edition 1962.

Industrial Diamond Information Bureau: *Hardness and hardness testing, 1937-1961.* London, the Bureau, second edition 1962.

Industrial Diamond Information Bureau: *Truing of grinding wheels, 1937-1961.* London, the Bureau, second edition 1962.

CHAPTER 14

BIBLIOGRAPHICAL CONTROL 2
ABSTRACTING AND INDEXING SERVICES, REVIEWS

ABSTRACTING AND INDEXING SERVICES: The mechanical engineer, as any other engineer or technologist, has only a limited amount of time available for reading. During this period it is impossible for him to do more than scan the main journals covering his field. The British engineer may for example regularly scan *Engineer, Engineering, Engineers' digest* and *Journal of mechanical engineering science,* while his American counterpart may peruse *Mechanical engineering, Machine design* and the various sections of *ASME transactions.* These journals, while containing much topical and significant material, will only account for a tiny proportion of all the articles of possible relevance to any one mechanical engineer. The remainder will be scattered through a large number of journals, any one of which may deal with either i) engineering in general; ii) mechanical engineering in general; iii) a specific division of mechanical engineering, *eg* welding or tribology; or iv) a subject field or industry not directly related to mechanical engineering but in which mechanical principles are adopted and mechanical engineers employed, *eg* the agricultural and food industries.

The extent of this scatter of mechanical engineering articles through periodicals can be gauged from the survey carried out by Peter Clague which indicated that in 1962 *Engineering index* included 2,230 abstracts of mechanical engineering articles from 370 different periodicals, *Applied mechanics reviews* contained 2,587 abstracts from 385 titles and *British technology index* included 1,094 references from 164 sources.[1]

Not only is it impossible for any individual to scan all of these journals; it will also not normally be possible for any one library or information service to acquire the titles. Excellent coverage of the literature of mechanical engineering can, however, be achieved by using abstracting and indexing services. In this way the engineer can utilize the limited reading time he has available to the best advantage.

An abstract is a summary of an article or document, accompanied by an adequate bibliographical reference to enable the article or document to be traced. Abstract journals are compilations of abstracts of articles currently being published in a wide range of journals, in addition to abstracts of other documents such as patent specifications and technical reports. The abstracts are brought together into subject groups which are then arranged either alphabetically under subject headings, as in *Engineering index*, or are classified according to a specially devised scheme as in *Applied mechanics reviews*. In addition to abstracting journals which consist solely of the abstracts, many other journals such as *Vaccum* and *Wear* contain abstracts as a regular feature. Both types of service are included in the listings given later in this chapter.

Abstracts are usually either indicative, annotations which merely indicate the scope and level of the original, or informative, summaries which actually present in a highly condensed form the most significant information disclosed in the original document.

Indexing services give less information about the subject content of the articles they list, in that they simply arrange bibliographical references to periodical articles (author, title of article, reference to journals, etc) under a series of alphabetical subject headings or classification symbols. Whereas with abstracting services the decision whether to consult the original can be based on the often detailed information given in the abstract, with indexing services selection is limited by the specificity of the subject heading provided by the indexer and the relevance of the title of the article provided by the author. Indexing services such as *British technology index* (BTI) and *Applied science and technology index* (ASTI) are, however, invaluable in that they get their entries into print with a minimum of delay. The time lag between the publication of an article and its registration in BTI is about seven weeks and with ASTI the delay is about fourteen weeks. Delays of up to ten months are common with abstracting services, and thus indexing services are particularly valuable for their early notifications of new developments.

Another application of indexing services is their use in picking up articles in fringe fields. The specialised abstracting services will cover only those journals which obviously relate to their own fields of

251

technology. An abstracting service covering metallurgy would not normally pick up an article on a metallurgical topic which was published in a journal covering agricultural engineering or space technology; both articles may nevertheless be of interest to the metallurgist. These articles would be indexed by BTI and ASTI, as both these services cover all important British and American technical journals respectively, regardless of their subject.

Abstracting and indexing services can be exploited in several ways:

i) Current awareness or alerting: they offer the only satisfactory method of keeping abreast of current developments in a specified field.

ii) Retrospective searching: abstracting and indexing services are the tools which enable engineers and librarians to search the literature over a given period to ascertain the state of a particular art, perhaps as a preliminary to the undertaking of a research project. This subject approach is made possible by consulting the subject indexes which are issued with each completed volume of most abstracting services.

iii) These services enable a searcher to trace a specific article of which only partial details are known or remembered, eg an enquirer may recall the author's name and the subject of an important paper he saw two or three years ago, but he may not recall the title of the journal in which the paper appeared. In this instance the paper could be traced through the author index of the abstracting service covering that particular field.

iv) Abstracting and indexing services are the most convenient source of reference to papers in the less accessible foreign languages. Slightly less than half the literature of science and technology is published in languages other than English. Only a relatively small percentage of engineers have even a reading knowledge of, for instance, Russian, and still fewer of Polish and Japanese. English language abstracts of papers in these languages, however, appear in *Engineering index* and *Applied mechanics reviews,* thus enabling the English speaking engineer to acquire knowledge of developments in other countries. In addition to services such as those mentioned above which cover the literature of all countries, other services provide English language abstracts of the technical literature of a particular country or group of countries, eg the Clearinghouse for Federal Scientific and Technical

Information's *USSR scientific abstracts* and *East European abstracts*. It is possible, if, for instance, the abstracting service provides long informative abstracts, that the information given in the English language abstract will be sufficient for the engineer's needs, but more often than not the abstract will serve as the basis on which to decide whether a full translation of the original should be obtained.

There are several published guides which enable us to identify abstracting and indexing services covering a particular discipline or industry. The most recent guide is the International Federation for Documentation's *Abstracting services. Vol 1: Science and technology*. The Hague, second edition 1969. An alphabetical listing of titles gives detailed information about each service and this is supplemented by classified, alphabetical subject and country indexes. Another comprehensive guide, which gives detailed information on 1,855 services published throughout the world, including number of abstracts published per year, frequency, types of index published, subject coverage, etc, is the National Federation of Scientific Abstracting and Indexing Services' *Guide to the world's abstracting and indexing services in science and technology*, Washington DC, 1963. A classified listing of services is followed by an alphabetical title listing and subject and country indexes. Detailed descriptive annotations on abstracting and indexing services are included in A J Walford's *Guide to reference material. Vol 1 Science and technology*, London, Library Association, second edition 1966. An alphabetical listing of services is given under the heading abstracting and indexing services in *Ulrich's international periodicals directory*. New York, Bowker, thirteenth edition 1969-1970, and a useful check-list of services in the English language is the National Lending Library for Science and Technology's *KWIC index to the English language abstracting and indexing publications currently being received by the National lending library*, Boston Spa, second edition 1967. Detailed information on new services is given as a regular feature in the International Federation for Documentation's monthly *FID news bulletin*, available from FID, 7 Hofweg, The Hague, Netherlands.

SERVICES COVERING PHYSICS

Papers and documents covering mechanics, fluid mechanics, thermodynamics and other physical subjects are abstracted in *Physics abstracts (Science abstracts, series A)*, which commenced publication in 1898 and which is issued twice a month by the Institution of Electrical Engineers (Savoy Place, London WC2) in association with other bodies including the American Institute of Physics, the Institute of Physics and the Physical Society. Coverage includes 1,100 journals published throughout the world, in addition to selected conference proceedings, reports and British and American patent specifications, and arrangement is under a broad subject classification. Subject and author indexes to *Physics abstracts* are published twice a year. The current awareness tool *Current papers in physics* is also a fortnightly publication issued by the institution, and this lists only the titles and bibliographical references of the documents abstracted in *Physics abstracts* under the same classified arrangement. *Bulletin signalétique*, comprising 36 individual sections, each covering a particular discipline, has been published in France under government sponsorship by the Centre de Documentation, Centre Nationale de la Recherche Scientifique (15 Quai Anatole-France, Paris 7) since 1940. Coverage is world-wide and the abstracts are of the brief indicative type. Section 130 covers general physics, physical mathematics, mechanics, acoustics, optics and thermodynamics while section 140 covers electricity. *Physikalische berichte* has been published each month by Deutsche Physikalische Gesellschaften EV (Burgplatz 1, 33 Braunschweig, Germany) since 1920. Coverage is world-wide, abstracts are in German, and arrangement is under broad subject with sections covering mechanics, thermodynamics and strength of materials. Detailed English language abstracts of articles from over 100 Russian periodicals covering physics are published in *Physics express*, a monthly which has been issued by International Physics Index (1909 Park Avenue, New York, NY 10035) since 1958.

SERVICES CONFINED TO MECHANICAL ENGINEERING

Abstracting services whose coverage is limited specifically to mechanical engineering are:

Japan science review. Mechanical and electrical engineering. 1954-. Q. Joint Publication Committee of the Engineering Society, Academic Press of Japan, Akaishi Dai-ichi Building, 60 Kanda-Jimbo-cho, 1-

chome, Chiyoda-ku, Tokyo, Japan. English language abstracts of Japanese technical literature.

Soviet abstracts. Mechanics. M. Ministry of Technology, TIL, Block A, Station Square House, St Mary Cray, Orpington, Kent. Available in two editions, *Titles only edition,* or *Selected abstracts edition.* The abstracts section includes selected abstracts translated from the Russian language abstracting service *Referativnyi zhurnal. Mekhanika.* Ceased Publication December 1968.

Technisches zentralblatt. Abteilung Maschinenwesen. 1952-. M. Institut fur Dokumentation der Deutschen Akademie der Wissenschaften zu Berlin, Akademie-Verlag, Leipziger Strasse 3-4, Berlin w8, Germany. Informative German language abstracts from the world's journal literature covering all aspects of mechanical engineering. Also includes detailed book reviews and notifications of American patent specifications.

SERVICES COVERING ENGINEERING IN GENERAL

The most comprehensive coverage in terms of English language abstracts of mechanical engineering literature is given by three services which cover technology in general, *Engineering index* (EI), *Applied science and technology index* (ASTI) and *British technology index* (BTI), and one covering applied mechanics, *Applied mechanics reviews* (AMR). Additional coverage of the foreign literature can be achieved by using:

a) foreign language abstracting services such as the relevant sections of *Bulletin signalétique;*

b) other general English language services covering the world's literature or the literature of a particular country, *eg Science citation index* or *Hungarian technical abstracts;* and

c) the specifically mechanical engineering services mentioned above.

Engineering index (EI), which is published by Engineering Index Inc (345 East 47th Street, New York, NY 10017), includes about 40,000 abstracts per year of the documents which are acquired by the Engineering Societies Library. EI offers the best coverage of any English language abstracting service covering mechanical engineering. EI's total coverage extends over all fields of technology and includes 3,500 journals published throughout the world and selected conference pro-

255

ceedings, books and patent specifications. Clague has estimated that in the 1962 volume of EI 72 percent of the abstracts were from English language journals and 19 percent from German, with Russian, French and Polish being the next most abstracted languages.[1] EI is available in a variety of forms. An annual volume has been published since 1885; its arrangement of abstracts is under alphabetical subject headings such as bearings, drilling machines, machine tools etc, some of which are further subdivided. Numerous *see also* references to related subjects facilitate generic searching and each annual volume is issued with an author index. A monthly version of the index, which follows the same arrangement as the annual volume, has been available since 1962, and a weekly card service of unit abstracts, which are available within 301 subject categories, was introduced in 1928. EI's latest development has been the introduction in 1969 of the COMPENDEX service (Computerised Engineering Index) which makes the monthly index available on magnetic tape.

Applied mechanics reviews (AMR) is a monthly critical review of the world literature covering applied mechanics and related engineering subjects, which has been published by the American Society of Mechanical Engineers (345 East 47th Street, New York, NY 10017) since 1948. In addition to long critical abstracts each issue of AMR includes an extensive review article covering the state of the art of a particular topic. Arrangement of the abstracts is under fifty five subject divisions which are in turn organised into six broad classes—mechanics of solids; fluid mechanics; heat etc, and the 10,000 abstracts published each year relate to the articles appearing in over 850 journals and also to selected books, conference proceedings and reports. Coverage of the lesser known foreign language journals is particularly good, but Clague notes that the better known foreign journals are more thoroughly abstracted by EI.[1] Each annual volume of AMR includes author and subject indexes.

Applied science and technology index (ASTI), published each month by H W Wilson Co (950 University Avenue, New York), commenced its life as *Industrial arts index* in 1913. The present title was adopted in 1957 when the original publication was split into ASTI and *Business periodicals index*. ASTI is a listing of titles of periodical articles, together with their bibliographical references, arranged under a series

256

of alphabetical subject headings, *eg* pumps; pumps, centrifugal; pumps, turbine; welding, welding equipment, some of which are further subdivided by sub-topic or country. Cumulated issues of ASTI are published each quarter and an annual volume is also included with the service. References are provided between related subjects, but ASTI lacks an author index. Approximately 100,000 entries to articles appearing in 230 English language journals, of which only about twenty are published in the UK, are included each year, but it must be noted that each article may be indexed under several subject headings.

Excellent coverage of British technical journals is offered by *British technology index* (BTI), a monthly which has been published by the Library Association (7 Ridgmount Street, Store Street, London WC1) since 1962. BTI is an alphabetical subject index covering some 400 British technical journals, but its subject entries are much more specific than those of ASTI, examples of typical BTI entries being: pumps, centrifugal: cavitation: effect of air content; and welding, oxy-acetylene: equipment: safety. Each article is only indexed under one heading, as alternative subject approaches to any one article are catered for by an integrated network of references. BTI's principle of specific entry makes it a very efficient tool for retrieving information on highly specific topics, but generic searching is decidedly easier under the broader headings of ASTI. BTI includes references to about 30,000 articles each year.

An additional service similar in its coverage to the above four publications is section 890 of *Bulletin signalétique—Sciences de l'ingenieur,* which extends over the engineering sciences including such subjects as automation, lubrication, heat engines, space technology and aeronautics. Other general services covering science and technology include:

Current contents—a series of weekly publications issued by the Institute for Scientific Information (325 Chestnut Street, Philadelphia, Pennsylvania 19106) each of which presents reduced reproductions of the actual contents pages of about 700 English language and foreign journals devoted to a related group of sciences, very soon after their publication. Individual sections are available for *Engineering and technology* and *Physical sciences.* Author indexes are included in each weekly issue.

257

9

Current index of scientific and technical literature. An alphabetical fortnightly subject index published by CCM Information Services Inc (866 3rd Avenue, New York, NY 10022), covering 2,100 English language and foreign journals on all aspects of pure and applied science, in addition to monographs and US government reports.

Referativnyi zhurnal. The world's most comprehensive abstracting service, published by VINITI (Lyubertsy-6, Oktiabr'skii Prospekt 403, Moscow, USSR), and currently available in twenty four separate sections. Coverage is international, and in addition to over 20,000 journals, monographs, theses and patent specifications are abstracted. Bibliographical information is given for each article in the language of the original following the Russian entry. In addition to those covering mechanical engineering, mechanics, and metallurgy, other sections are available covering automatic control, automobile engineering, physics, metrology, refrigeration, mechanical handling, instrumentation, space technology, power plant, piping, shipbuilding, air transportation, railway engineering.

Science citation index (SCI). An unique computer produced index published quarterly and cumulated annually by the Institute for Scientific Information (325 Chestnut St, Philadelphia, Pennsylvania 19106), which enables the searcher to move forward in time from a paper of known interest to the other published papers which have subsequently cited the original paper. SCI thus enables a watch to be made on the applications and developments of the ideas disclosed in any particular paper. Cited papers are arranged alphabetically by their author and each is accompanied by bibliographical details of citing papers. Coverage is of 2,000 journals relating to all fields of science and technology. SCI is particularly useful for searching in interdisciplinary areas where the papers may be scattered through a number of journals devoted to different subject fields. Abstracting services are not well suited to catering for these situations. The institute's ASCA (Automatic Subject Citation Alert) service will automatically keep a client informed of newly published papers relevant to his subject field.

English language abstracting services covering the technical literature of a particular country include:

Abstracts of Romanian technical literature. M. Central Institute for Documentation, Raisnul 30 Decembrie, Str Cosmonautilor, nr 27/29,

Bucharest, Rumania. Informative English language abstracts of articles from the Rumanian technical press. Sections cover metallurgy, mechanical engineering, plastics.

Australian science index. 1957-. M. Commonwealth Scientific and Industrial Research Organisation, 314 Albert Street, East Melbourne C2, Australia. An index of articles published in Australian scientific and technical periodicals.

East European scientific abstracts. 2 per month. Joint Publications Research Service, Clearinghouse for Federal Scientific and Technical Information, Springfield, Virginia 22151, USA. Separate series, each including informative English language abstracts, cover *Cybernetics, computers and automation technology, Engineering and equipment, Metals and metallurgy,* and *Physics and mathematics.*

Hungarian technical abstracts. 1949-. Q. Hungarian Central Library and Centre for Documentation, Box 8, Budapest 8, Hungary. English language abstracts from the Hungarian technical press and translated contents lists of Hungarian technical periodicals. Subject coverage includes physics, mechanical engineering, properties of materials, control engineering, etc.

Indsoc list of current scientific literature. 1954-. 2 per month. INDSOC (Indian National Scientific Documentation Centre), National Physical Laboratory, New Delhi, India. A classified index to articles appearing in the world's technical periodicals.

Monthly index of Russian accessions. 1948-1969. M. 'A record of the publications in the Russian language issued in and outside the Soviet Union that are currently received in the Library of Congress and a group of cooperating libraries.' Included translated contents lists of Russian technical and scientific journals and a classified subject index to their contents. Ceased publication in May 1969.

Polish scientific periodicals—contents. 10 per year. Export and Import Enterprise 'Ruch', ul Wronia, Warsaw, Poland. Reproduced contents pages, many in English, of 150 Polish periodicals covering science and technology.

Polish technical abstracts. 1951-. Q. Centralny Instytut Dokumentacji Naukowo-Techniczenj, Warsaw, Aleja Niepodleglosci 188. English language abstracts of articles from Polish technical periodicals. Sections cover metallurgy and mechanical engineering.

Power express: the comprehensive digest of current Russian literature dealing with power topics. 1961-. 10 per year. International Physical Index Inc, 1909 Park Avenue, New York, NY 10035. Informative English language abstracts with subject coverage including atomic power, rotating machinery, heat exchange, combustion and fuel, automation, etc.

Summaries of articles from the French press. Q. Centre National du Commerce Extérieur, 10 Av D'Iéna, Paris 16.

USSR scientific abstracts. 2 per month. Joint Publications Research Service, Clearinghouse for Federal Scientific and Technical Information, Springfield, Virginia 22151, USA. Separate series, each including informative English language abstracts, cover *Cybernetics, computers and automation technology, Engineering and equipment, Metals and metallurgy* and *Physics and mathematics.*

In addition to the services listed above which cover either science or engineering, or both, or which cover the entire field of mechanical engineering, a number of more specialised services exist to cater for specific divisions of mechanical engineering or to cover particular industries. These services will be used by organisations with specialised interests to supplement the coverage given by the four most important abstracting and indexing services—*Engineering index, Applied mechanics reviews, Applied science and technology index* and *British technology index.*

AERONAUTICAL AND AEROSPACE ENGINEERING

International aerospace abstracts. 1961-. 2 per month. American Institute of Aeronautics and Astronautics, Technical Information Center, 750 3rd Avenue, New York, NY 10017.

Pacific aerospace library uniterm index to periodicals. 1944-. 14 per year. American Institute of Aeronautics and Astronautics. A subject index to papers published in 300 English language journals covering aeronautical and astronautical engineering and related sciences.

ALUMINIUM

Aluminium abstracts. 1963-. BI-M. Centre International de Development de l'Aluminium, 6 Avenue Bertie-Albrecht, Paris 8e.

260

APPLIED HEAT

BCURA monthly bulletin. 1937-. M. British Coal Utilisation Research Association, Randalls Road, Leatherhead, Surrey.

Bulletin signalétique. Section 730—Combustibles. Énergie thermique. 1940-. Q. Centre de Documentation, Centre Nationale de la Recherche Scientifique, 15 quai Anatole-France, Paris 7.

Bulletin synoptique de documentation thermique. 1962-. M. Institut Français de Combustibles et de l'Énergie, 3 rue Henri Heine, Paris 16°.

Fuel abstracts and current titles: a monthly summary of world literature on all technical aspects of fuel and power. 1960-. M. Institute of Fuel, 18 Devonshire Street, Portland Place, London W1.

Gas abstracts. 1945-. M. Institute of Gas Technology, 3424 South State Street, Chicago, Illinois 60616, USA.

Heat bibliography. 1948-. Annual. Fluids Group, National Engineering Laboratory, East Kilbride, Glasgow, Scotland. Literature references are arranged under alphabetical subject headings and include references to books, with, in some cases, reference being made to reviews of individual volumes.

AUTOMATION AND INSTRUMENTATION

Automation express: digest of current Russian literature dealing with automation topics. 1958-. 10 per year. International Physical Index Inc, 1909 Park Avenue, New York, NY 10035.

Computer and control abstracts (Science abstracts ser C). 1966-. M. Institution of Electrical Engineers, Savoy Place, London WC2.

Current papers on computers and control. 1966-. M. Institution of Electrical Engineers. Bibliographical references only to papers abstracted in the above publication.

SIRA abstracts and reviews. 1946-. M. Taylor & Francis Ltd, Red Lion Court, Fleet Street, London EC4, for British Scientific Instrument Research Association.

AUTOMOBILE ENGINEERING

Automobile abstracts. 1955-. M. Motor Industry Research Association, Nuneaton, Warwickshire.

Bulletin mensuel de documentation. 1945-. M. Union Technique de l'Automobile, 2 rue de Presbourg, Paris 8.

COPPER

Copper abstracts. 1951-. Bi-M. Copper Development Association, 55 South Audley Street, London W1.

CORROSION

Bibliographic surveys of corrosion. 1945-. Irreg. National Association of Corrosion Engineers, 1061 M & M Building, Houston 2, Texas, USA.

Corrosion abstracts: abstracts of the world's literature on corrosion and corrosion mitigation. 1962-. Bi-M. National Association of Corrosion Engineers.

ENGINE DESIGN AND APPLICATION

Abstracts from technical and patent publications. 1947-. M. British Internal Combustion Research Institute, 111/112 Buckingham Avenue, Slough, Buckinghamshire.

Bulletin signalétique. Section 730—Combustibles. Énergie thermique. 1940-. Q. Centre de Documentation, Centre Nationale de la Recherche Scientifique, 15 quai Anatole-France, Paris 7.

ENGINEERING DESIGN

Engineering design abstracts. Formerly published by Enfield College of Technology in association with the Institution of Engineering Designers—now incorporated as a regular feature in *Engineering designer* published by Institution of Engineering Designers, 36 Portland Place, London W1. Includes current articles and publications of interest to individuals concerned with mechanical engineering design.

Ergonomics abstracts. 1969-. Q. Ergonomics Information Analysis Centre, University of Birmingham, Birmingham.

FABRICATION AND JOINING

Bibliographical bulletin for welding and allied processes. 1949-. Q. International Institute of Welding, 32 bd de la Chapelle, Paris 18.

FLUID POWER AND PNEUMATICS

British hydromechanics research association bulletin. 1948-. 5 per year. British Hydromechanics Research Association, Cranfield, Bedford, England. Includes abstracts covering hydraulic pumps and turbines;

fluid mechanics; pipes and fittings; fluid meters; seals, bearings, lubrication, etc.

Fluid power abstracts. 1970-. M. British Hydromechanics Research Association.

Fluid sealing abstracts. 1970-. M. British Hydromechanics Research Association. Includes abstracts covering static, general dynamic, and reciprocating rotary seals and sealing materials.

Fluidics feedback: a current information guide. 1967-. M. British Hydromechanics Research Association. Includes literature abstracts and review articles.

Index bibliographique du vide. 1966-. BI-M. International Union for Vacuum Science, Technique and Applications, 30 av de la Renaissance, Brussels 4, Belgium.

HEATING AND VENTILATING ENGINEERING

Thermal abstracts. BI-M. Heating and Ventilating Research Association, Old Bracknell Lane, Bracknell, Berkshire.

INSTRUMENTATION see AUTOMATION AND INSTRUMENTATION

IRON AND STEEL

BCIRA abstracts of foundry literature. 1969-. BI-M. British Cast Iron Research Association, Alvechurch, Birmingham.

BSCRA abstracts. 1952-. BI-M. British Steel Castings Research Association, East Bank Road, Sheffield 12.

ABTICS service. Iron and Steel Institute, 4 Grosvenor Gardens, London SW1. See page 23.

Stahl und eisen. Zeitschriften und bucherschau. 1969-. BI-M. Verein Deutscher Eissenhuttenleute, Breite Str 27, Düsseldorf, Germany.

LEAD

Lead abstracts: a review of recent technical literature on the uses of lead and its products. 1951-. M. Lead Development Association, 34 Berkeley Square, London W1.

LOCOMOTIVE ENGINEERING

British railways research department. Monthly review of technical literature. 1951-. Research Department, British Railways, London Road, Derby.

Railway research and engineering news. 1964-. 3 per year. Railway Research Index Division, 1 Burg Eijssenstr, Wijnandsrade, Netherlands. Pt A: Abstracts from Russian railway periodicals; Pt B: Abstracts from other periodicals published throughout the world.

Selection of international railway documentation. 1961-. M. International Union of Railways, 14/16 rue Jean Rey, Paris 15ᵉ.

LUBRICATION see TRIBOLOGY

MACHINE TOOLS see WORKSHOP TECHNOLOGY

MARINE ENGINEERING
Journal of abstracts, British Ship Research Association. 1946-. M. British Ship Research Association, Wallsend Research Station, Wallsend, Northumberland.

Marine engineering and shipbuilding abstracts. Appears as a regular monthly feature in *Transactions of the Institute of Marine Engineers.* Institute of Marine Engineers, Memorial Buildings, 76 Mark Lane, London EC3.

MATERIALS SCIENCE AND ENGINEERING
References on fatigue. 1950-. Annual. American Society for Testing and Materials, 1916 Race Street, Philadelphia 3, Pennsylvania, USA.

METAL FINISHING
Metal cleaning bibliographical abstracts. 1943-. Irreg. American Society for Testing and Materials, 1916 Race Street, Philadelphia 3, Pennsylvania, USA.

Metal finishing abstracts. 1959-. BI-M. Finishing Publications Ltd, 17 Grosvenor Road, Hampton Hill, Middlesex.

METALLURGY
Bulletin signalétique. Section 740—Métaux. Métallurgie. 1940-. M. Centre de Documentation, Centre Nationale de la Recherche Scientifique, 15 quai Anatole-France, Paris 7.

Metals abstracts. 1966-. M. Institute of Metals, 17 Belgrave Square, London SW1. Supersedes *ASM review of metal literature,* 1944-1967 and *Metallurgical abstracts,* 1908-1967.

264

Montanwissenschaftliche literaturberichte. Abteilung B: Metallurgie. 1955-. M. Akademie-Verlag, GMBH, Leipziger Str 3-4, Berlin W8, Germany (DDR).

NICKEL

Nickel bulletin: monthly abstracts of recent technical literature. 1928-. M. International Nickel Co Inc, 67 Wall Street, New York, NY 10005.

NON-FERROUS METALS

BNF bulletin. 1921-. M. British Non-Ferrous Metals Research Association, Euston Road, London NW1.

NUCLEAR ENGINEERING

Atomindex. 1959-. 2 per month. International Atomic Energy Agency, Karntner Ring 11-13, A-1010, Vienna 1, Austria. Bibliographical references to report literature, conference proceedings, microforms, etc.

Bulletin signalétique. Section 150. Physiques et technologie nucléaires. 1940-. M. Centre de Documentation, Centre Nationale de la Recherche Scientifique, 15 quai Anatole-France, Paris 7.

Nuclear science abstracts. 1947-. 2 per month. Division of Technical Information Services, United States Atomic Energy Commission, PO Box 62, Oak Ridge, Tennessee, USA.

Nuclear science abstracts of Japan. 1962-. M. Japan Atomic Energy Research Institute, Division of Technical Information, Tokai-mura, Naka-gun, Ibaraki-ken 319-11, Japan.

Transatom bulletin. 1960-. M. Euratom, Transatom Service, 51 rue Billiard, Brussels, Belgium. References to translations into Western European languages from Russian, Eastern European and Japanese technical literature.

PLASTICS

Plastics abstracts. 1959-. Weekly. Plastics Investigations, 31 Canonsfield Road, Welwyn, Hertfordshire. Abstracts of British patent specifications.

RAPRA abstracts. 1967-. M. Rubber and Plastics Research Association, Shawbury, Shrewsbury, Shropshire.

PLATINUM
Platinum metals review. 1957-. Q. Johnson, Matthey & Co Ltd, Hatton Garden, London EC1.

POWDER METALLURGY
Metal powder report. 1946-. M. Powder Metallurgy Ltd, Paramount House, 75 Uxbridge Road, London W5.

PRODUCTION ENGINEERING
PERA bulletin. 1948-. M. Production Engineering Research Association, Melton Mowbray, Leicestershire.
Production technology: abstracts and reports from Eastern Europe. Q. Machine Tool Industry Research Association, Hulley Road, Macclesfield, Cheshire.

QUALITY CONTROL AND MEASUREMENT
Quality control and applied statistics: international literature digest service. 1958-. M. Executive Sciences Institute Inc, Whippany, New Jersey 07981, USA.
Reliability abstracts and technical reviews. 1967-. M. National Aeronautics and Space Administration, Reliability and Quality Assurance Office, Washington, DC 20546, USA.

REFRIGERATION ENGINEERING
Cryogenic information report. 1963-. 16 per year. Technical Economics Associate, Estes Park, Colorado 80517, USA. Information on new equipment and processes and new cryogenic applications with a limited number of literature abstracts.
International institute of refrigeration. Bulletin. 1920-. BI-M. Institut International du Froid, 177 bd Malesherbes, Paris 17e. The bibliographical references which appeared in the Bulletin for the period 1953-1960 have been cumulated and published by Pergamon Press as *Bibliographical guide to refrigeration.* 1962.

SPRINGS
Spring journal. 1946-. Q. Spring Research Association, Doncaster Street, Sheffield.

THERMODYNAMICS see APPLIED HEAT

TRIBOLOGY

British Hydromechanics Research Association bulletin. 1948-. 5 per year. British Hydromechanics Research Association, Cranfield, Bedford, England. Includes abstracts covering bearings and lubrication.

Tribos. 1968-. M. British Hydromechanics Research Association. Contents cover lubrication fundamentals, friction, wear, lubricants, materials, applied lubrication.

Wear includes a monthly section devoted to *Systematic abstracts of current literature.*

WELDING see FABRICATION AND JOINING

WORKSHOP TECHNOLOGY

Industrial diamond review. 1941-. M. Industrial Diamond Information Bureau, Arundel House, Kirkby Street, London ECI.

Machine tool industry research association. Library bulletin. 1968-. M. Machine Tool Industry Research Association, Hulley Road, Macclesfield, Cheshire.

PERA bulletin. 1948-. M. Production Engineering Research Association, Melton Mowbray, Leicestershire.

ZINC

Zinc abstracts: a monthly review of recent technical literature on the uses of zinc and its products. 1947-. M. Zinc Development Association, 34 Berkeley Square, London W1.

REVIEWS

Reviews are compilations which either summarise or evaluate developments in a highly specific field as documented in the literature over a given period of time. Reviews are invaluable to the engineer who wishes to know the state of a particular art. The review will consist of a narrative survey of developments, followed by a bibliography listing the items summarised or evaluated in the narrative. Examples of reviews which are essentially critical and evaluative are those published in *Reviews of modern physics,* while an example of a review which simply aims at providing an indicative summary of developments in a particular field by referring the reader to the numerous published documents is ASME's annual review of lubrication literature published in the *Journal of lubrication technology*. The 1967 survey listed no less than 467 references.

Reviews are published either as i) individual articles in primary journals; ii) collections of reviews in specialised reviewing journals; or iii) annual or irregular surveys in volumes entitled *Advances in . . .* or *Progress in . . .*

Individual review articles often appear in *Journal of mechanical engineering science*, which in 1969 adopted the policy of publishing occasional review articles of between 5,000 and 6,000 words in length. Examples of subjects covered in 1969 include ' Review of theories of metal removal in grinding ' and ' Review of the interaction of creep and fatigue '. Other journals which regularly feature review articles include *Chartered mechanical engineer*, *eg* 'A review of bending, forming and shearing ', and *Wear*, *eg* ' Packing rub effect in rotating machinery: a state of the art review '.

Specialised reviewing journals covering all aspects of physics include, in addition to the American Institute of Physics' *Reviews of modern physics*, the Institute of Physics and the Physical Society's bi-monthly *Reports on progress in physics*, which also contains critical and evaluative literature surveys. *Applied mechanics reviews* features one review article in each issue, a recent example of which, ' Effect of sound and vibrations on heat transfer ' covered over 250 references. 120 papers which have appeared as the review articles in *Applied mechanics reviews* over a number of years have been conveniently presented in a single volume as *Applied mechanics surveys* (Washington, Spartan Books, 1966) under the editorship of H N Abramson. The surveys are arranged in subject groupings under the headings ' mechanics of solids ', ' dynamics ', ' mechanics of fluids ' and ' heat '. Other papers covering such topics as numerical analysis or automatic and adaptive control have been grouped under a general section. Review articles published in either primary journals or in specialized reviewing journals can usually be traced through abstracting services.

The annual or irregularly published surveys of progress normally include about six critical articles in each volume. Examples of such series of relevance to mechanical engineers include:

Advances in applied mechanics. 1948-. Irregular at intervals of 2/3 years. New York, Academic Press.

Advances in computers. 1960-. Annual. New York, Academic Press.

Advances in heat transfer. 1964-. Annual. New York, Academic Press.

Advances in materials research. 1967-. Irreg. New York, Wiley.

Advances in nuclear science and technology. 1962-. Biennial. New York, Academic Press.

Progress in applied materials research. 1959-. Irreg. London, Heywood.

Progress in automation. 1960-. Irreg. London, Butterworth.

Progress in control engineering. 1962-. Irreg. London, Heywood.

Progress in cryogenics. 1959-. Irreg. London, Heywood.

Progress in materials science, incorporating *Progress in metal physics.* 1949-. Irreg. Oxford, Pergamon Press. These highly specialized reviews have recently been issued as separates in addition to their publication as collections of four or five reviews in bound format.

Progress in metallurgical technology. 1960-. Irreg. London, Iliffe, for Institution of Metallurgists.

A further category of publication also entitled *Advances in . . .* are published, not as critical reviews, but as collections of papers on a central theme which have been presented at an annual conference or a symposium. Such publications are often issued in a continuing series and in a standard format, examples are:

Advances in automobile engineering. 1963-. Annual. Oxford, Pergamon Press. Volumes of papers presented within the Cranfield International Symposium Series at the Advanced School of Automobile Engineering. Each volume deals with a specific topic—vehicle ride; automatic transmission; noise and vibration; etc.

Advances in cryogenic engineering. 1960-. Annual. New York, Plenum Press. Proceedings of the annual Cryogenic Engineering Conference.

Advances in machine tool design and research. 1962-. Irreg. Oxford, Pergamon Press. Each volume includes papers presented at the International Machine Tool Design and Research Conference.

REFERENCE

1 Clague, P: *The coverage of the periodical literature of mechanical engineering by published abstracts journals.* Thesis submitted for the Fellowship of the Library Association. 1967.

CHAPTER 15

THE FOREIGN LANGUAGE PROBLEM AND SOME SOLUTIONS

A survey carried out recently by the National Lending Library for Science and Technology revealed the following facts:

1 Fifty per cent of the world's scientific and technical literature is published in languages other than English.

2 British scientists and technologists are frequently confronted by a paper in a foreign language which they wish to read but are unable to because of language difficulties.

3 British scientists appear to be largely unaware of the existing services which would help them to overcome the language barrier.[1]

The most formidable section of this language barrier is that between the Russian and English languages. Russian is now perhaps the second most productive language of papers on science and technology, but only a very small percentage of those scientists and technologists whose native language is English are capable of reading Russian. Translation charges in the UK from the Russian language vary between £4 10s ($11.00) to £9 10s ($23.00) per thousand words, and thus the cost of producing an English version of a Russian paper can be extremely high. Facilities do exist, however, for making generally available those translations which have already been made of Russian and other originals by firms and organisations in Great Britain and abroad in fulfilment of previous requests. These facilities are location indexes of translations, which refer an enquirer to an organisation willing to make a translation available, and translation pools—actual collections of translations.

The Aslib *Commonwealth index to unpublished translations* was established in the United Kingdom in 1951 following a recommendation by the Royal Society Scientific Information Conference, 1948. This index contains locations for over 200,000 translations and this number is being added to at the rate of 12,000 new locations each year. Over 300 organisations throughout Great Britain and the Commonwealth notify Aslib of their newly completed translations, and in addition the index includes entries from the various published listings of translations such as *Translations register index* (see below). The largest collection of translations in the British Isles is maintained at the National Lending Library for Science and Technology. Translations are acquired from many sources, both at home and abroad, and the collection currently contains over 120,000 separate items. Co-operation exists between NLL and Aslib, and NLL consults the Aslib index before answering any enquiry for a translation. The combined NLL/Aslib satisfaction figure for such enquiries is now around fifteen per cent. Recent additions to the NLL loan collection of translations are listed in the monthly *NLL translations bulletin*. An invaluable feature of NLL's translation services is the Russian article *ad hoc* translation scheme, under which an enquirer in the United Kingdom can request, at no cost to himself providing he agrees to edit a draft, a translation of any Russian paper published within the previous two years on an aspect of science, technology or the social sciences from any journal not scheduled for cover-to-cover translation.

The National Translations Center (NTC) was established in the United States in 1953 at the John Crerar Public Library (35 West 33rd Street, Chicago, Illinois 60616), under the auspices of the Special Libraries Association and with financial assistance from the National Science Foundation. The John Crerar Public Library assumed complete responsibility for the center in 1969, again with National Science Foundation support. NTC receives copies of translations from over 200 private organisations throughout the United States and Canada, and the total number of translations held currently is in excess of 136,000. New additions to the collection are listed in *Translations register-index*, which is published twice a month with a cumulated quarterly index. NTC also published a *Consolidated index of translations into English* in 1969. Translations which are made by the United States Atomic Energy Commission, the Department of Defense, the National Aeronautics and Space Administration and other governmental agencies, and which are released to the public by the Clearinghouse for Federal Scientific and Technical Information are listed in *United States government research and development reports*.

The European Translations Centre (ETC) is maintained at the Technological University, Delft, Netherlands, and is both a translations pool and a translations index. In addition to holding copies of over 150,000 translations, from Russian and other eastern European languages into western European languages, donated by organisations from 17 countries, locations are available for 750,000 additional translations. A *List of translations notified to ETC,* arranged by subject, is published each month and this attempts to include those translations which have not been cited in other published lists. The *World index of scientific translations,* arranged under title of the original journal and then chronologically by article, is the most comprehensive listing of translations currently available. This computer produced journal appears quarterly, cumulates annually, and includes notifications of translations made to ETC, in addition to details of additional translations gleaned from other published lists.

Translations of books are listed annually in Unesco's *Index translationum: international bibliography of translations,* vol 1-, 1948-. Previous to this date the work had been published quarterly by the International Institute of Intellectual Cooperation from 1932 to 1940. The bibliography is arranged initially by the countries in which the translations are published, and then inside this arrangement by broad subject grouping. An author index is provided.

Two translation services covering particular industries are the British Iron and Steel Industry Translation Service (BISITS) (Iron and Steel Institute, 4 Grosvenor Gardens, London SW1) and Transatom (European Atomic Energy Agency, CID-Transatom, 51/53 rue Belliard, Brussels, Belgium). BISITS was established in 1957 as a cooperative service between the major British steel companies, and has since become the main source of English translations of literature on iron and steel production and related topics. Important articles are selected for translation, and limited numbers are reproduced for the benefit of contributors and others. Weekly lists of BISITS translations and subject bibliographies of translations are available on request from the Iron and Steel Institute. From 1970 the service has also undertaken to provide private translations of specific articles or documents on request. The Transatom service produces translations of articles and documents relating to nuclear science and technology from Russian, eastern European, and the other less accessible languages into western European languages, and notification of these is given each month in *Transatom bulletin.*

Cover-to-cover translations from Russian into English have been produced in increasing numbers since 1949. These have all the advantages of conventional journal publication in that they can be circulated, stored, indexed and abstracted, and in addition an engineer can keep himself informed of technological progress in the Soviet Union by regularly scanning a cover-to-cover journal relating to his field. A real shortcoming of this form, however, is that often these translations take many months to appear; delays of up to twelve months are not unusual, and consequently they can never completely stem the demand for *ad hoc* translations. Cover-to-cover journals on specific subjects relating to mechanical engineering are included in the subject directory of periodicals, chapter 10. The *NLL translations bulletin* from time to time publishes a complete list of those available. The last such list, published in the June 1969 issue, extended over 283 titles and included a KWIC (key-word in context) index to their subject range. Many of these journals are produced under government subsidy, in the United States under the auspices of the Clearinghouse for Federal Scientific and Technical Information, and in the United Kingdom under the sponsorship of the National Lending Library for Science and Technology.

It is often necessary to obtain the services of a translator to cover an article of which an existing translation is not available. The following guides to translation services and lists of translators are useful in this context:

Millard, P: *Directory of technical and scientific translators and services.* London, Crosby Lockwood, 1968. Covers the British Isles only and gives names of individual translators with brief details of their language qualifications and subject interests; details of translation bureaux with descriptive notes on their services; subject and language indexes.

Kaiser, F E: *Translators and translations: services and sources in science and technology.* New York, Special Libraries Association, second edition 1965. Covers private translators and translating firms in the United States and Canada, but in addition gives international

274

lists of translation pools and information services for translations and also of bibliographies of translations.

. A register of translators which includes over 200 names with information on the language and subject interests of each is maintained in the United Kingdom by Aslib. Similar indexes have also been compiled by large public technical libraries in the United Kingdom and the United States and these may be consulted without charge. Two published indexes of translators issued by professional bodies are:

Institute of Linguists: *Index of members of the translators' guild.* London, the Institute, 1969.

American Translators' Association: *Professional service directory.* New York, the Association, second edition 1969.

REFERENCE

1 Wood, D N: 'The foreign language problem facing scientists and technologists in the United Kingdom: report of a recent survey'. *Journal of documentation* 23(2), June 1967, 117-130.

INDEX

Abramson, H N 268
Abrasive engineering 157
Abrasive methods 157
Abridgement of specifications 166-168
Abstracting services 250-267
Abstracting services (FID) 253
Abstracts
defined 251
types 251
Abstracts of Romanian technical literature 258
ABTICS (Abstract and book title index card service) 23, 263
Academie Polonaise des Sciences. Bulletin. Serie des sciences techniques 112
Academy of Sciences of the USSR. Proceedings. Applied physics section 109
AC-Delco news 124
Acier/Stahl/Steel 134
Aciers speciaux 135
Acrow digest 140
Acta mechanica 110
Acta metallurgica 143
Acta Polytechnica Scandinavica. Mechanical engineering series 116
Acta technica 112
Acta technica CSAV 113
Adams, F D 204
Advancement of science 108
Advances in . . . series 268-269
Advisory Council for Scientific Policy 45
Advisory Council on Technology 45
Advisory Group for Aerospace Research and Development (AGARD) 222, 248
Aeronautical dictionary 204
Aeronautical engineering
abstracting and indexing services 260
data sheets 181
periodicals 118-120
reference books 204-205
report literature 75, 184
research centres: UK 38-39, 41-42; USA 75-76, 77-78

Aeronautical engineering—*contd.*
societies: UK 22; USA 67
standards 174-175
trade associations: UK 59; USA 87
Aeronautical journal 118
Aeronautical quarterly 118
Aeronautical Sciences Data Unit 181
Aeroplane and Armament Experiment Establishment 41
Aeroplane directory of British aviation 205
Aerospace engineering
abstracting and indexing services 260
periodicals 118-120
reference books 204-205
report literature 75, 184
research centres: USA 75-76, 77-78
societies: UK 22; USA 67
sources of information 11
standards 174-175
trade associations: UK 59; USA 87
Aerospace facts and figures 205
Aerospace Industries Association of America 87
Aerospace structural metals handbook 225
Aerospace yearbook 205
AFS cast metals research journal 131
AIAA journal 118
Ainley D G 9, 10, 15
AIPE newsletter 149
Air BP 119
Air Conditioning and Refrigeration Institute 89, 91
standardization activities 173
Air conditioning, heating and refrigeration news 133, 217
Air conditioning, heating and ventilating 133
Air tables 214
Air University 75-76
Aircraft and missile design and maintenance handbook 204
Aircraft engines of the world 205
Aircraft Research Association 38-39
Airmec-AEI Ltd 54

American Society of Lubrication Engineers 68
standardization activities 173
American Society of Mechanical Engineers 64-65, 68
founded 64
periodical publications 100, 115-116, 245
preprints 100, 245
research activities 65
standardization activities 173
American Society of Naval Engineers 68
American Society of Tool and Manufacturing Engineers 69
standardization activities 173-174
American Standards Association 177
American Translators' Association 275
American Welding Society 70
standardization activities 174
American Zinc Institute 92
Ames Laboratory (NASA) 77
Anderson, E P 236
Anderson, H H 213
Annuaire de la presse française 106
Anthony L J 11
Anti-corrosion manual 210
Anti-corrosion methods and materials 125
Anti-Friction Bearings Manufacturers' Association 87
Anvil 114
Apex 117
Appliance manufacturer 152
Applied heat
abstracting and indexing services 261
periodicals 120-121
reference books 206
research centres: UK 55
Applied mechanics reviews 250, 255, 256, 268
Applied mechanics surveys 268
Applied plastics 150
Applied science and technology index 250-252, 255, 256-257
Archiv für das eisenhuttenwesen 134
Archiwum budowy maszyn 116

Archiwum mechaniki stosowanej 110
Argus 159
Arnell, A 198
Arnold Engineering Development Center (USAF) 73
Aro Inc 73
ASCA (Automatic subject citation alert) 258
Ash, L 93
ASHRAE guide and data book 173, 216, 233
ASHRAE journal 133
ASHRAE transactions 133
ASLE transactions 155
ASLIB 48-49, 185
founded 48
functions 48
research department 49
short courses 48
subject groups 48
translation activities 271-272, 275
Aslib booklist 241, 243
Aslib directory 51
ASM review of metal literature 264
ASM boiler and pressure vessel code 173
ASME handbook
1: metals properties 224
2: engineering tables 199
3: metal engineering design 225
4: metals engineering—processes 236
ASME power test codes 173
Assembly and fastener methods 128
Assembly engineering 128
Association of British Aluminium and Gold Bronze Powder (Flake) Manufacturers 59
Association of Consulting Engineers 43
Association of Hydraulic Equipment Manufacturers 61
Association of Iron and Steel Engineers 68
Association of Light Alloy Refiners and Smelters 59
Association of Shell Boilermakers 60
ASTM standards 172-173
ATB metallurgie 143
Atom 147

281

Graham, A K 223
Grant, H E 236
Grant-aided research associations 31-38
Gray, D W 195
Gray and Ductile Iron Founders' Society 88
Grazda, E E 197
Greenwood, D C 202
Grinding
  bibliographies 249
  periodicals 157-160
  trade associations: USA 90
Grinding Wheel Institute 90
Grits and grinds 160
Grytz, E 199
Guide to American directories 195
Guide to current British periodicals 105
Guide to metallurgical information 12
Guide to reference books 194
Guide to reference material 194, 253
Guide to scientific and technical periodicals 106
Guide to sources in space science and technology 11
Guide to special issues and indexes of periodicals 195
Guide to the literature of mathematics and physics 12
Guide to the world's abstracting and indexing services in science and technology 253
Guide to uncommon metals 227

H & VR catalogue of heating and directory of heating engineers 217
Haberer, I J 103
Habicht, F H 237
Haldane report 44, 45
Hall, N 223
Hampel, C A 227
Handbook of adhesives 212
Handbook of air conditioning, heating and ventilating 217
Handbook of air conditioning system design 216

Handbook of analytical design for wear 234
Handbook of applied instrumentation 207
Handbook of applied mathematics 197
Handbook of astronautical engineering 204
Handbook of automation, computation and controls 207
Handbook of barrel finishing 223
Handbook of brittle material design technology 222
Handbook of chemistry and physics 197
Handbook of compressed gases 213
Handbook of dimensional measurement 226
Handbook of electron beam welding 211
Handbook of engineering fundamentals 197
Handbook of engineering mechanics 202
Handbook of experimental stress analysis 222
Handbook of fastening and joining of metal parts 212
Handbook of fixture design 237
Handbook of fluid dynamics 214
Handbook of generalised gas dynamics 214
Handbook of heat transfer media 206
Handbook of heating, ventilating and air conditioning 216
Handbook of high vacuum engineering 214
Handbook of industrial electroplating 223
Handbook of industrial metrology 226
Handbook of instrumentation and controls 207
Handbook of lattice spacings and structures of metals and alloys 225
Handbook of mathematical tables 199
Handbook of mechanical wear 234
Handbook of metal powders 231
Handbook of noise control 226

289

10

Institute of Physics and the Physical Society 25
Fulmer Research Institute 21, 43
register of consultants 44
Institute of Refrigeration 25
Institute of Refrigeration proceedings 154
Institute of Road Transport Engineers journal and proceedings 123
Institute of Sheet Metal Engineering 24
Institute of Textile Technology 70
Institute of Welding 21
Institution of Electrical Engineers 21
Institution of Engineering Design 22
Institution of Engineering Inspection 25
Institution of Engineers and Shipbuilders in Scotland 23
Institution of Engineers and Shipbuilders in Scotland transactions 138
Institution of Engineers, Australia journal 113
Institution of Engineers, Australia mechanical and chemical transactions 116
Institution of Engineers (India) journal, mechanical engineering division 116
Institution of Heating and Ventilating Engineers 22
Institution of Heating and Ventilating Engineers journal 133
Institution of Locomotive Engineers 23
Institution of Locomotive Engineers journal 136
Institution of Mechanical Engineers
data activities 181
founded 19
library and information services 23-24
periodical publications 99, 115, 245
research activities 20
subject groups 20
Institution of Metallurgists 24
Institution of Mining and Metallurgy 21, 24

Institution of Mining Engineers 24
Institution of Nuclear Engineers 25
Institution of Plant Engineers 25
Institution of Production Engineers 21, 25
Institutions
UK 16-25; USA 64-70
Instrument and control engineering 122
Instrument engineer 123
Instrument manual 208
Instrument practice 122
Instrument Society of America 67
Instrument technology 122
Instrumentation
abstracting and indexing services 261
bibliographies 246
periodicals 121-123
reference books 206-208
research centres: UK 34, 35
societies: UK 22; USA 67
trade associations: UK 60
Instrumentation 123
Instruments and control systems 122
Insulation 134
Interdok 187
Interlab 54
Internal Combustion Engine Institute 89
Internal combustion engines
abstracting and indexing services 262
periodicals 126-127
reference books 210-211
research centres: UK 42, 43
trade associations: UK 61; USA 89
International aerospace abstracts 260
International Congress on Combustion Engines 179
International conversion tables 200
International critical tables 200
International encyclopedia of science 196
International Federation for Documentation 253
International Institute of Refrigeration 179

Letters patent 161
Lewis', H K library 243
Leybold vacuum handbook 214
Leyland journal 117
Library and information services
UK: national 48-51; research associations 32-38; research stations 40; societies and institutions 21-25
USA: federal government 73-84; non-federal 66-70, 93-95
Library of Congress
National Referral Center for Science and Technology 82-83
Science and Technology Division 83
Library of Congress catalog 241, 244
Liddell, D M 227
Lifting Equipment Manufacturers' Association 62
Light metal age 146
Light metals handbook 227
Light production engineering 152
Link-belt news 140
Lipson, C 222, 234
List of serials covered by members of the NFSAIS 105
List of translations notified to ETC 273
Little science, big science 15
Lloyds' register of ships 221
Lockheed horizons 120
Locomotive and Allied Manufacturers' Association 62
Locomotive engineering
abstracting and indexing services 263-264
periodicals 136
reference books 218
societies: UK 23; USA 68
trade associations: UK 62
Lubrication
abstracting and indexing services 267
bibliographies 248
periodicals 155-156
reference books 234-235
research centres: UK 40, 55; USA 72-73
reviews 268
societies: USA 68

Lubrication—*contd.*
standards 173
trade associations: UK 62; USA 90
Lubrication 156
Lubrication engineering 156

McBirnie, S C 220
MacGibbon's MOT orals and marine engineering knowledge 219
McGraw-Hill basic bibliography of science and technology 243
McGraw-Hill encyclopedia of science and technology 196
MacGregor, C W 234
Machine and tool blue book 158
Machine and tool directory 238
Machine design 137
Machine design and control 137
Machine design engineering 137
Machine devices and instrumentation 218
Machine moderne 158
Machine shop and engineering manufacture 158
Machine tool catalog file 190
Machine tool directory 238
Machine tool industry: UK 52
Machine Tool Industry Research Association 36-37, 52
Machine Tool Industry Research Association bulletin 267
Machine tool review 160
Machine tool specification manual 237
Machine Tool Trades Association Inc 57, 63
Machine tools
abstracting and indexing services 267
bibliographies 246
periodicals 122, 157-160
reference books 207, 236-238
research centres: UK 29, 36-37, 40-41
societies: USA 69
trade associations: UK 63; USA 91
Machinery
periodicals 152-153, 157-160
trade associations: UK 62; USA 90
Machinery 158

295

297